and all the other cowboy dads
who only like cow math.

Visit boredontherange.com
to see more!

For inquiries, please email
rangeboss@boredontherange.com

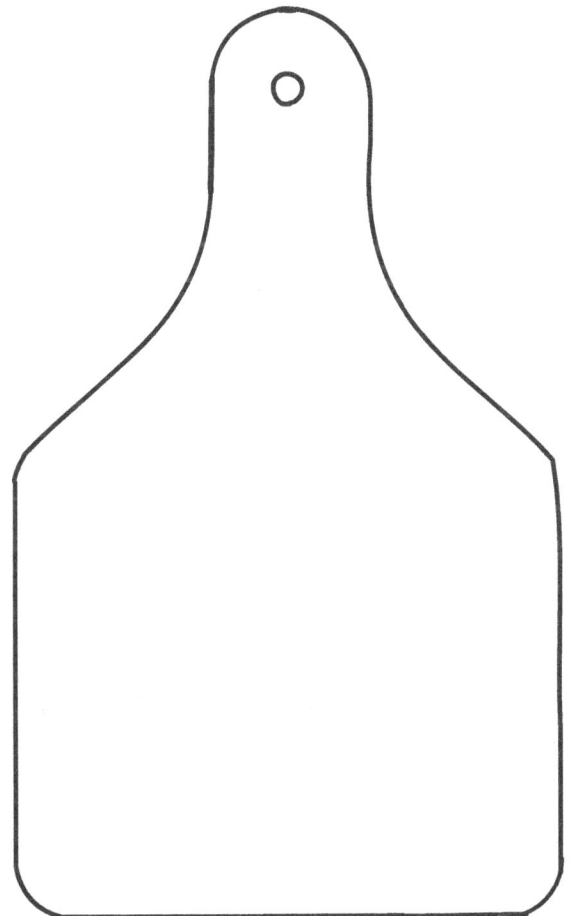

123

RANCH MATH

ONE

1

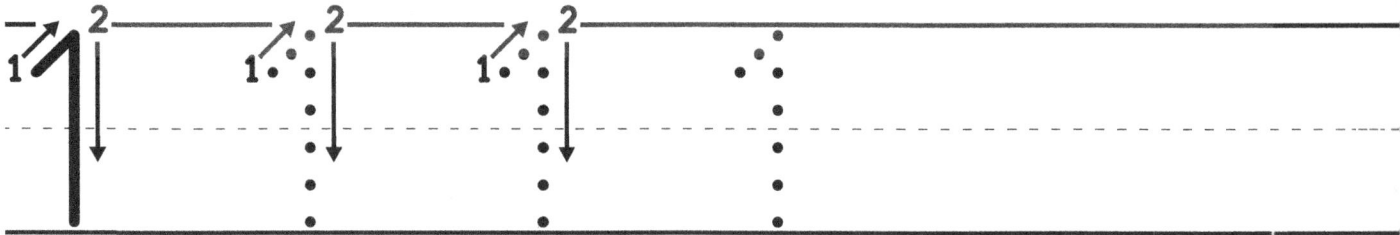

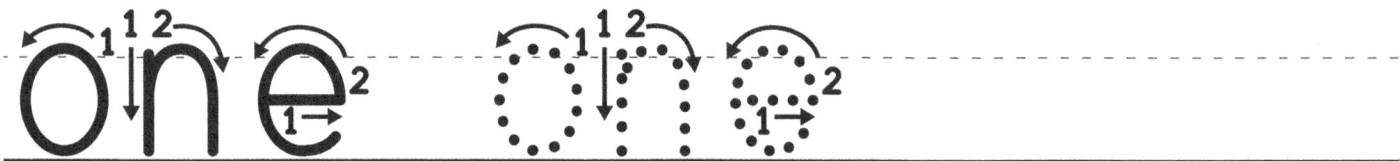

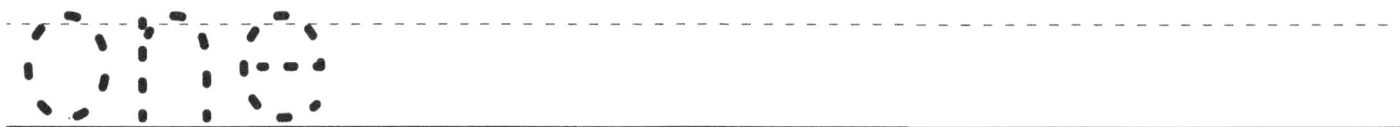

2

The dragger needs to
rope ONE calf. Use red
to color the calf he should
rope. Color the other
calves black.

TWO

2

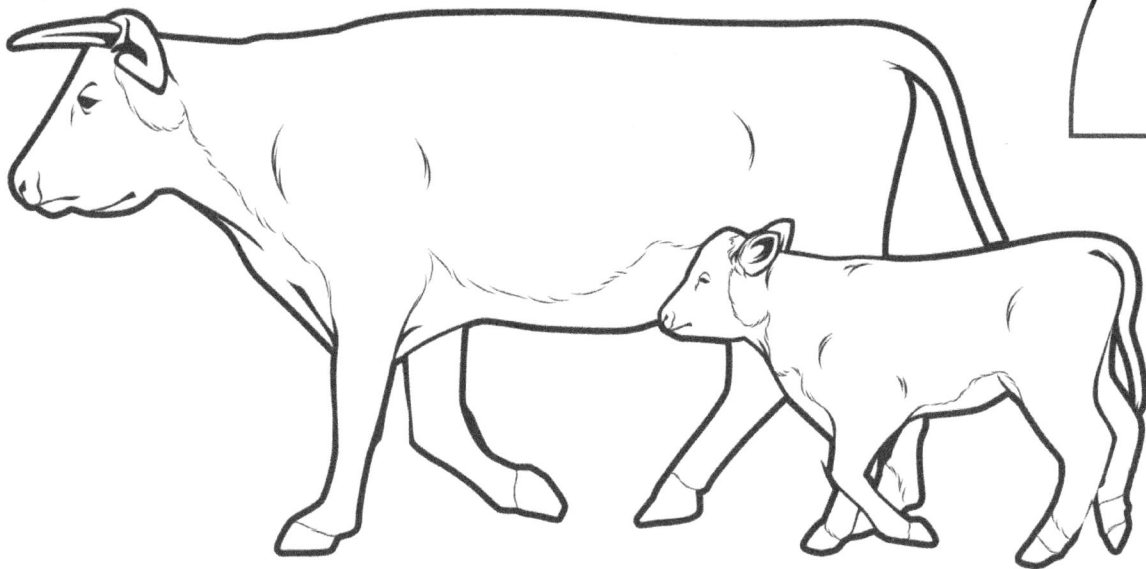

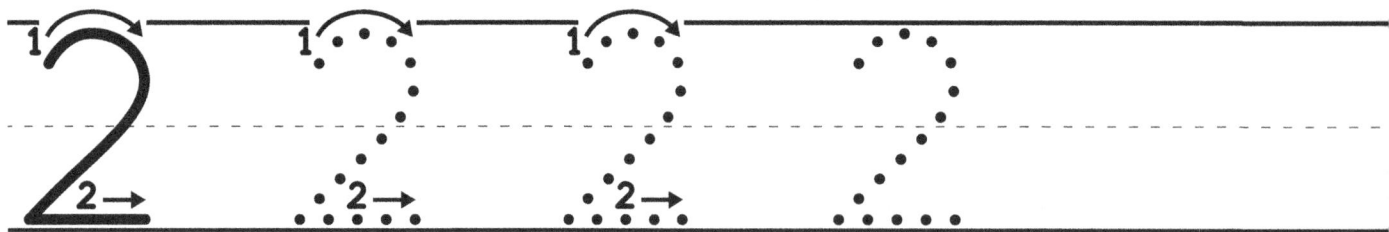

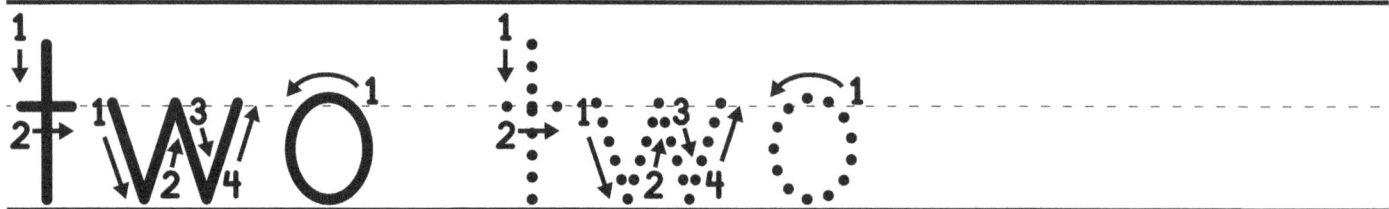

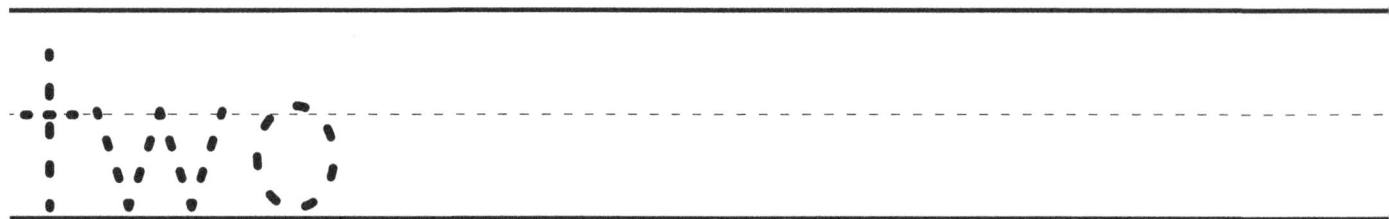

4

A cow and calf are called a pair because there are TWO. Cowboy Jackson and his horse, Rooster, need to move the pairs to a new pasture. Can you color all of the pairs in this picture?

THREE

3

Triangles have THREE sides. Each side of a teepee is a triangle.

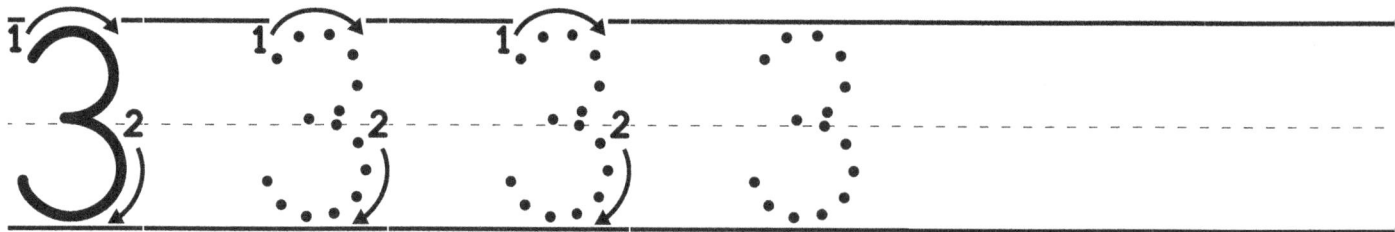

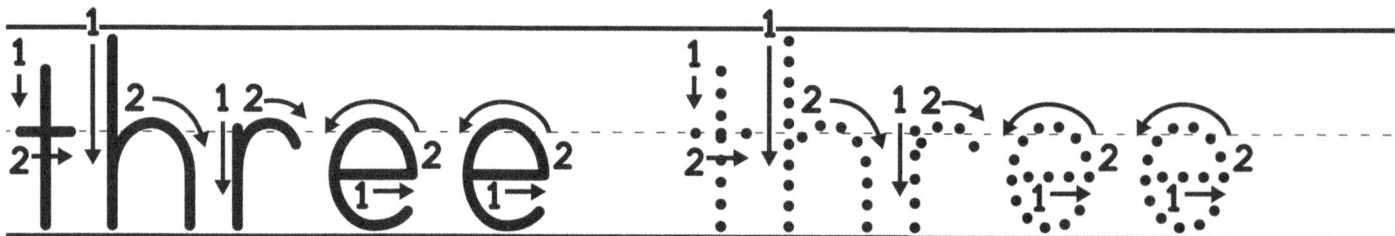

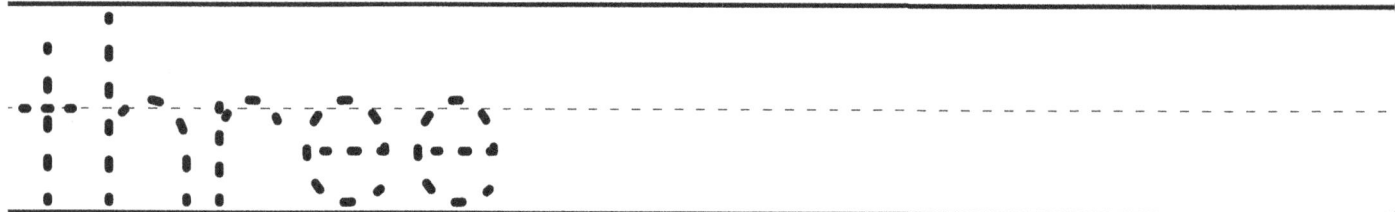

3 3 3 3

three three

three

Can you draw a
triangle teepee next to
the roping dummy?

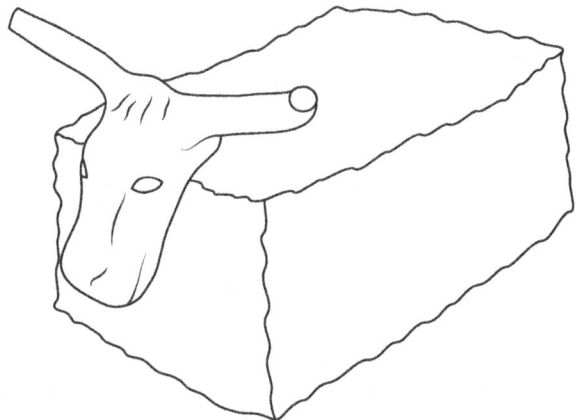

FOUR

4

There are FOUR horses grazing in the pasture. Color one horse bay, one horse sorrel, one horse gray, and one horse paint!

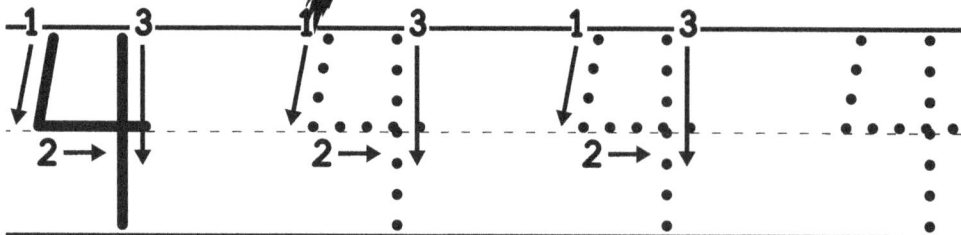

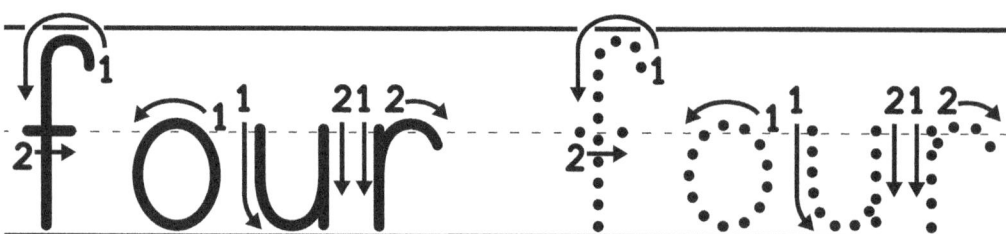

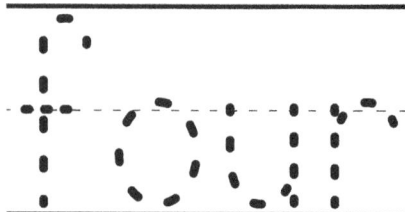

4 4 4 4

four four

four

FIVE

5

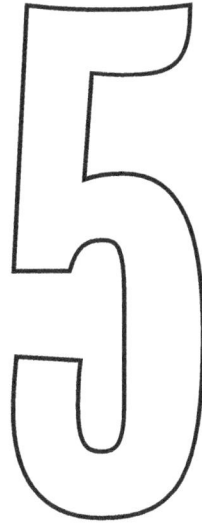

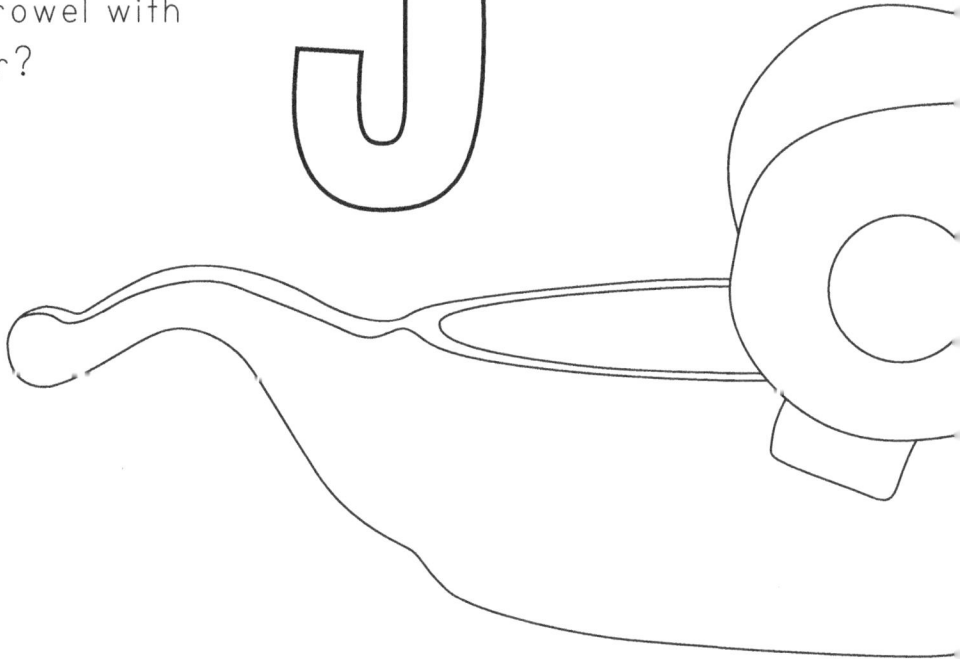

Oh, no! Cowboy Kevin lost a spur rowel. Can you draw a rowel with FIVE points on the spur?

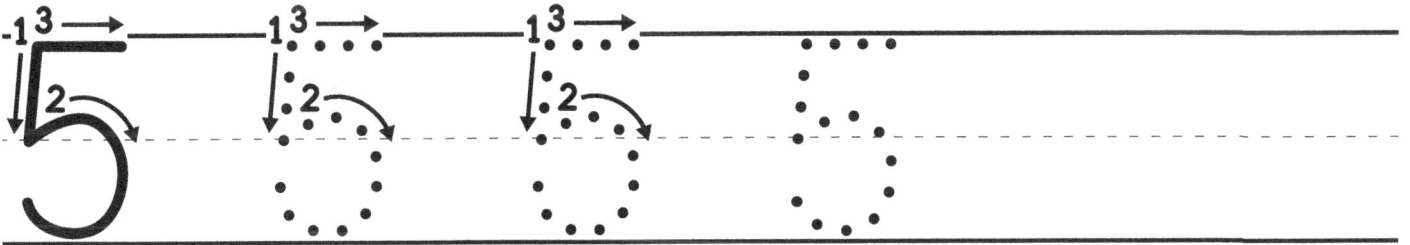

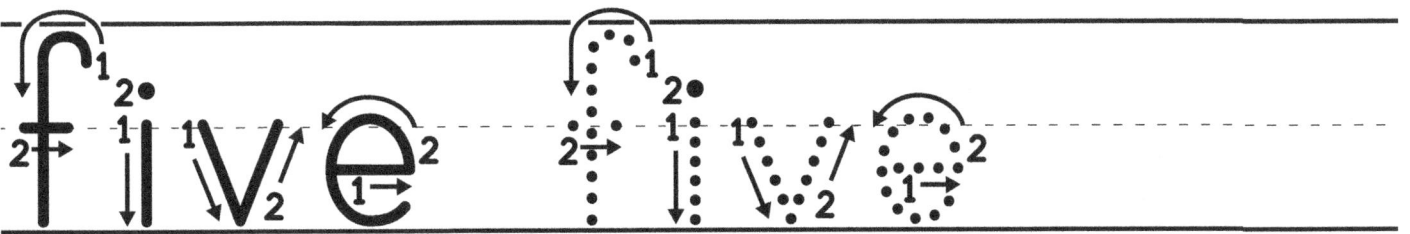

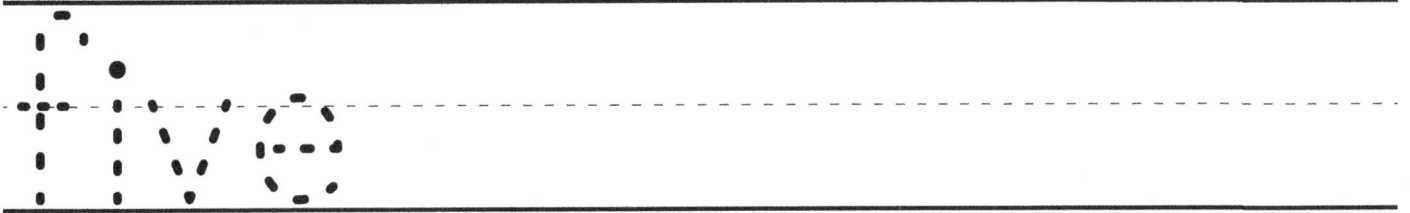

5 5 5 5

five five

five

SIX

Everett roped three calves on Saturday. Hilldo, his mighty horse, helped him rope three more calves. $3 + 3 = 6$. They roped a total of SIX calves!

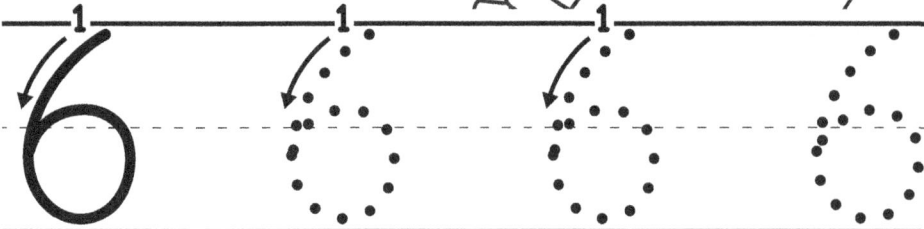

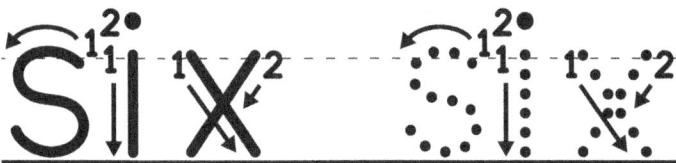

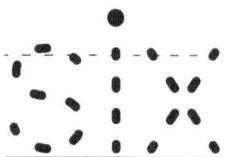

6

6 6 6 6

SIX SIX six

six

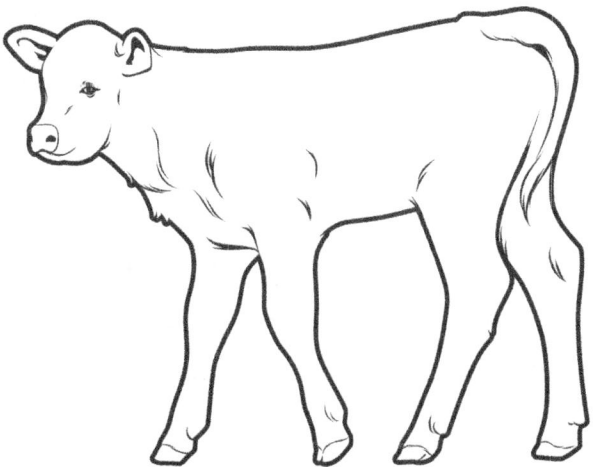

Four of the calves are steers and need blue tags. The remaining two are heifers and need gold tags.

Can you draw ear tags on the calves before you color them?

SEVEN

7

Cowboy Josh needs to drive seven T-posts in the fence. Can you draw SEVEN T-posts in the barbwire for him?

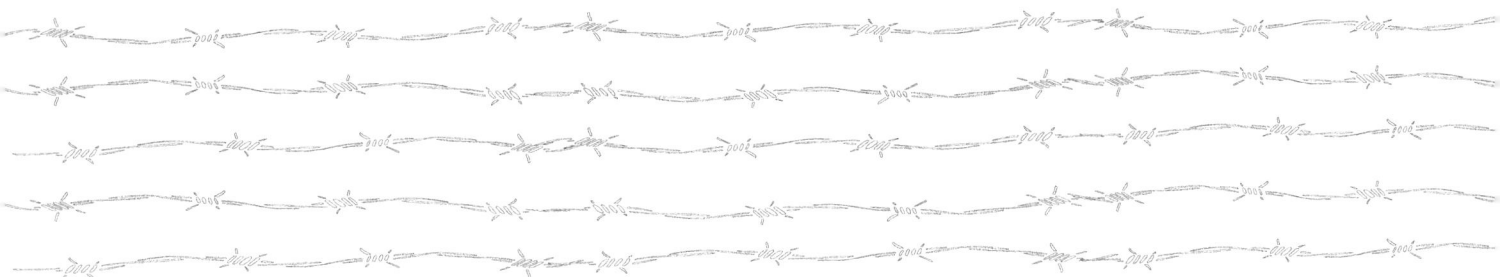

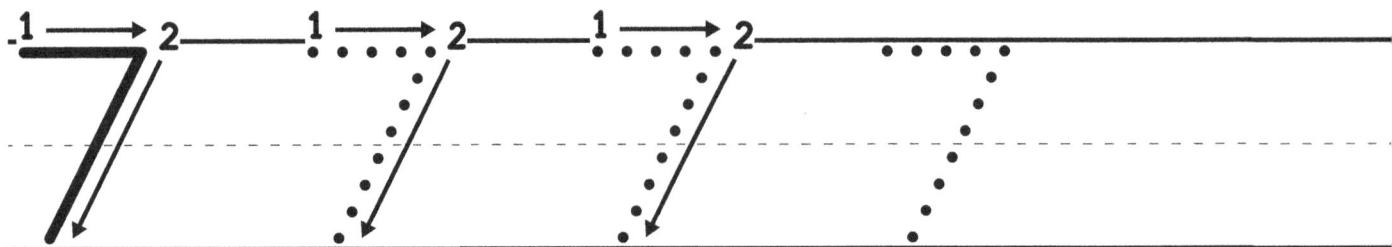

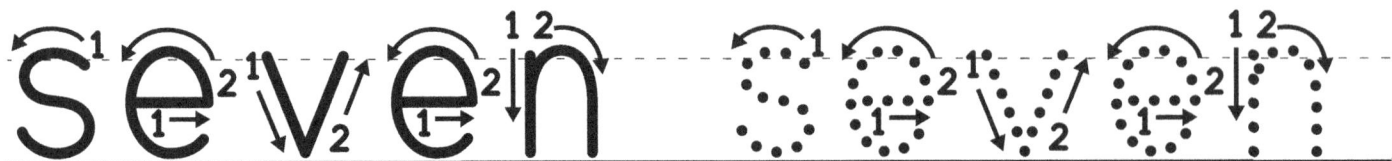

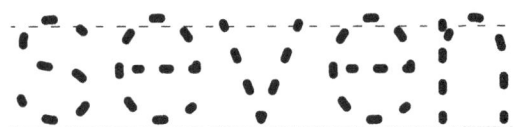

7 7 7 7

seven seven

seven

EIGHT

8

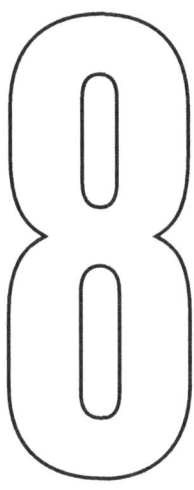

The cowboys have to stay on their bronc for EIGHT seconds to receive a score from the judges.

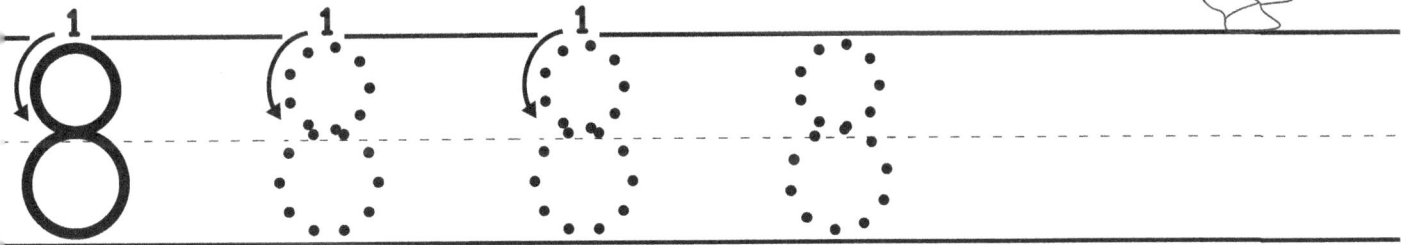

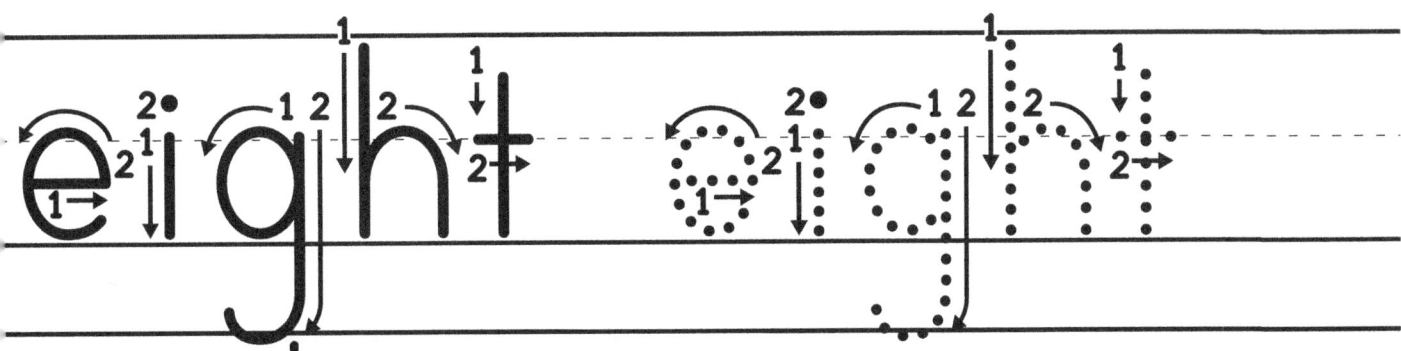

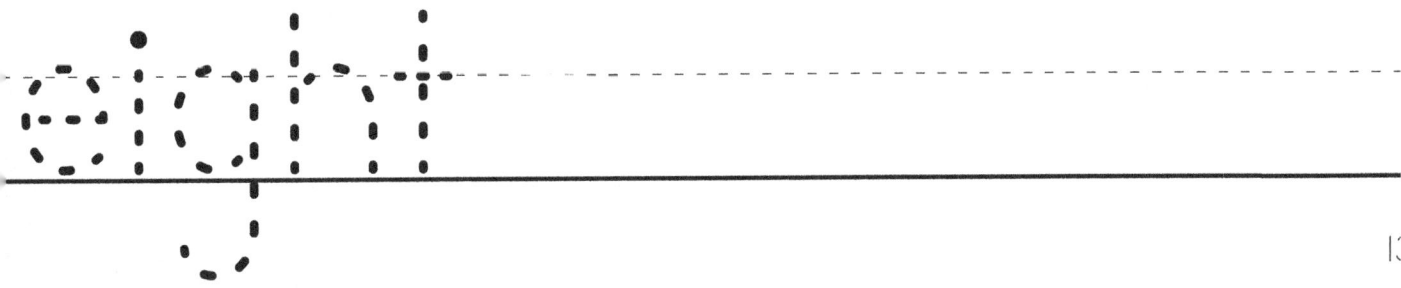

8 8 8 8

eight eight

eight

13

NINE

A round pen is a circle-shaped pen used for horse training and sometimes roping. Landon is training a cutting horse to sort cattle.

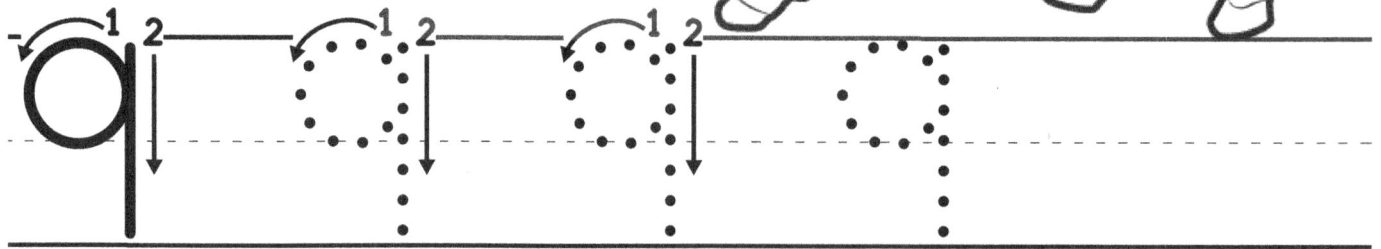

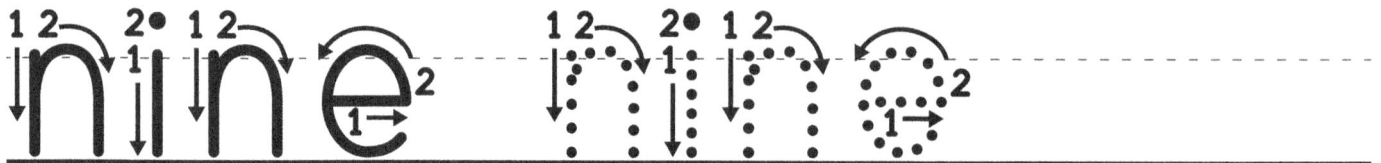

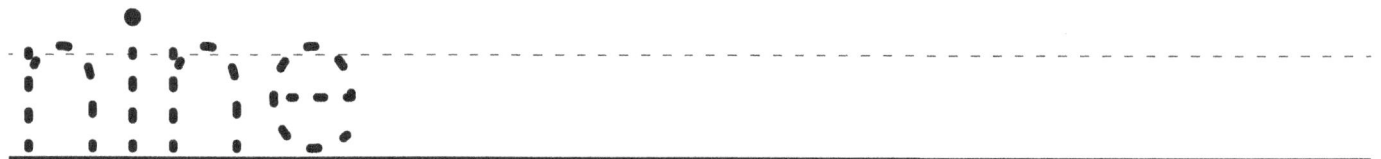

9

q q q q

nine nine

nine

Can you draw a round pen with NINE cows in it to
help Landon train his cutting horse?

TEN

10

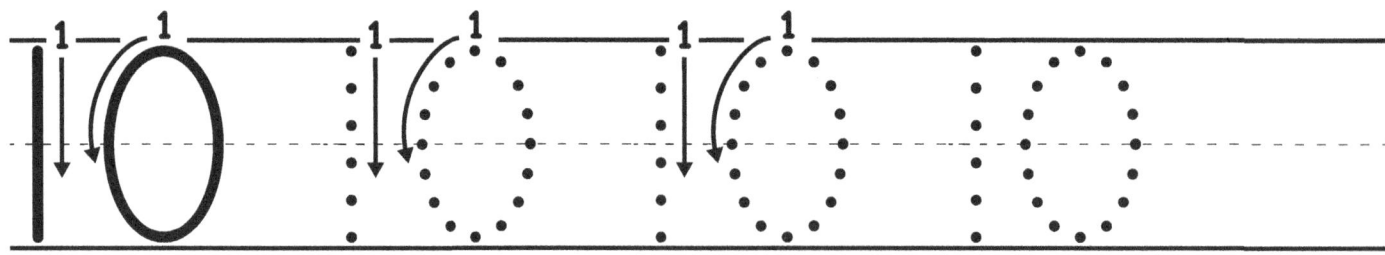

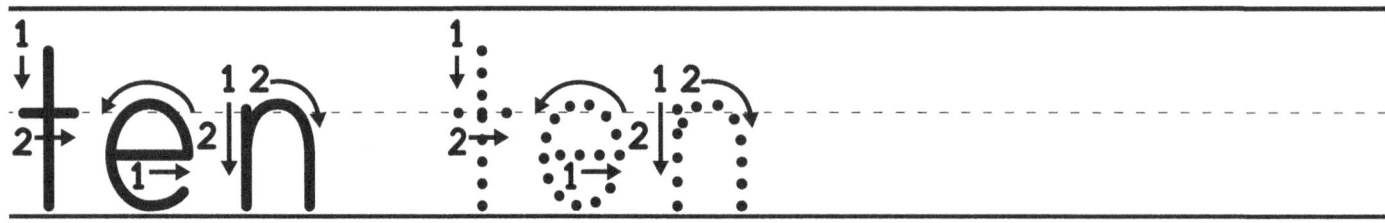

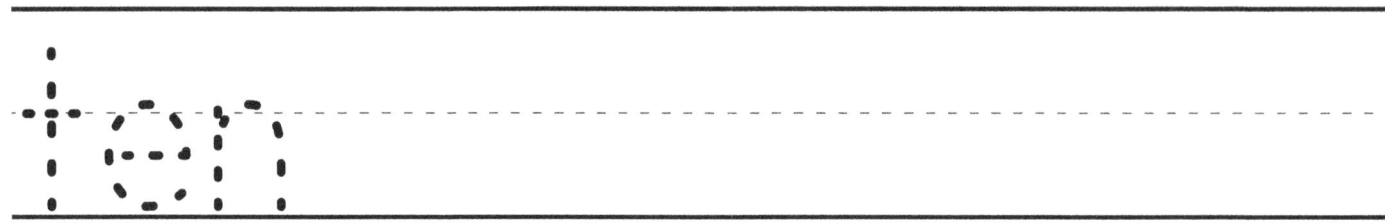

Ty needs TEN cc's of medicine in her vaccine gun. Can you color in the vaccine so she has enough orange medicine to doctor the cattle?

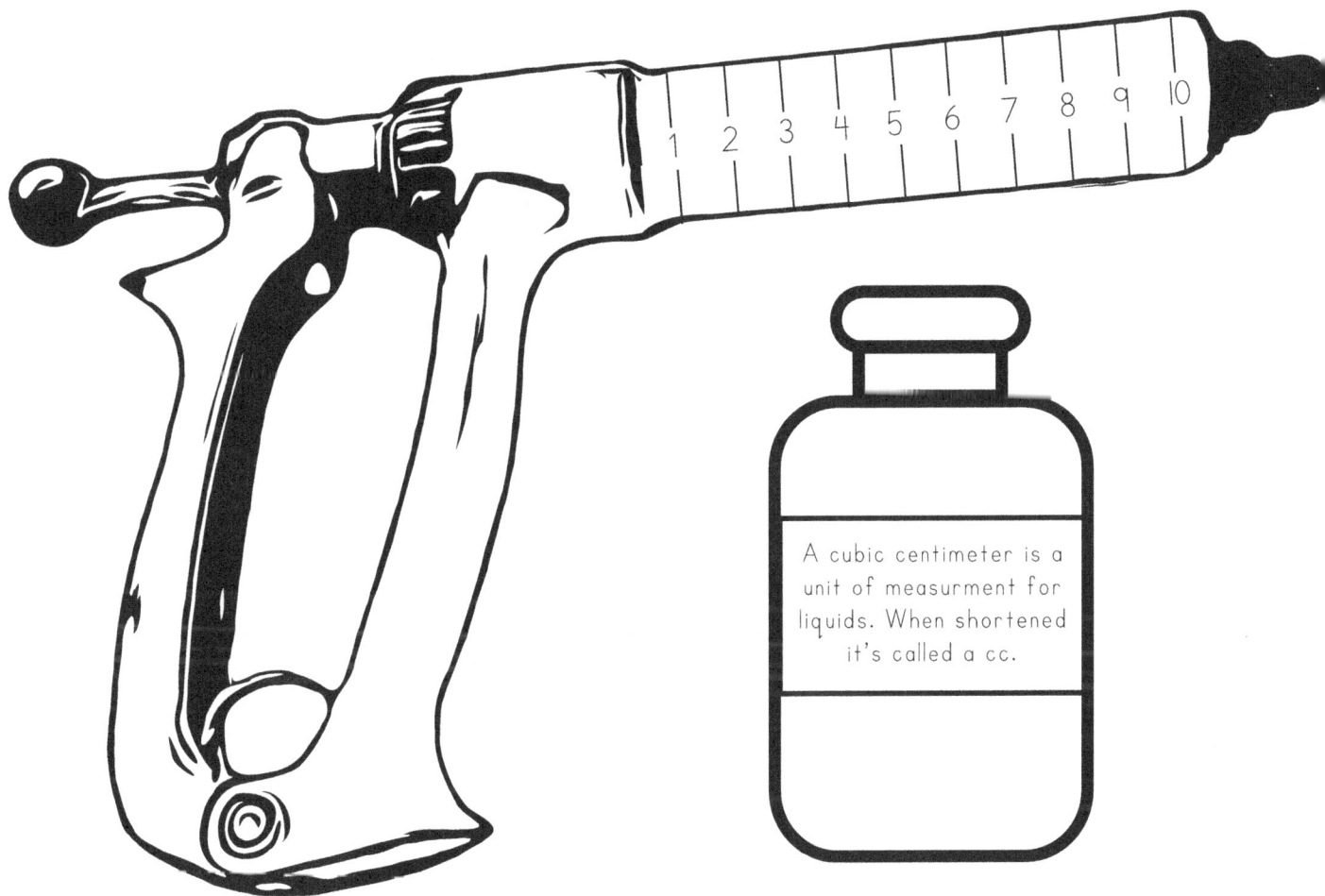

1 2 3 4 5 6 7 8 9 10

A cubic centimeter is a unit of measurment for liquids. When shortened it's called a cc.

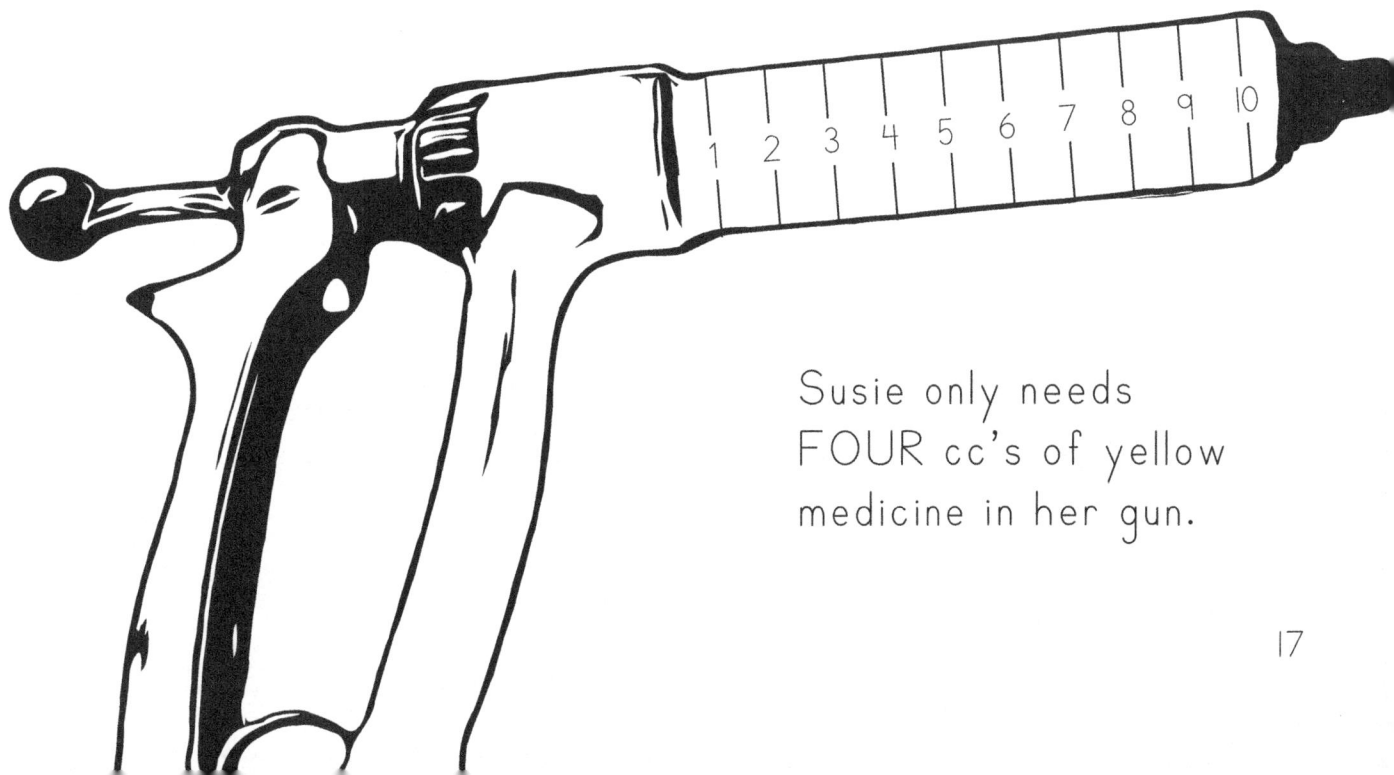

1 2 3 4 5 6 7 8 9 10

Susie only needs FOUR cc's of yellow medicine in her gun.

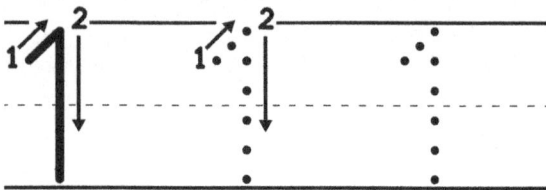

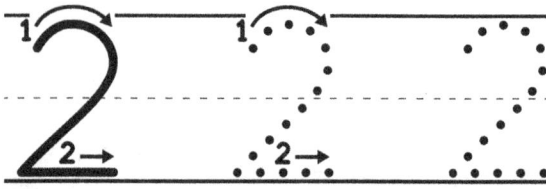

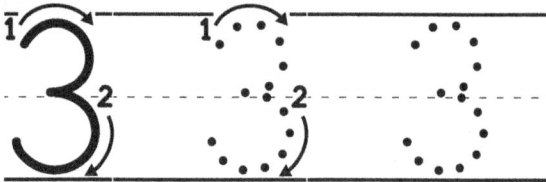

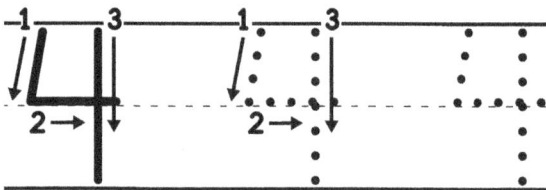

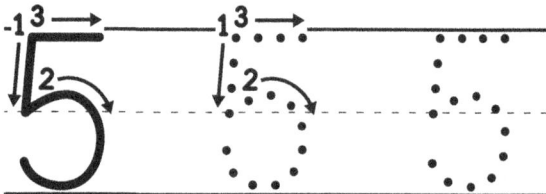

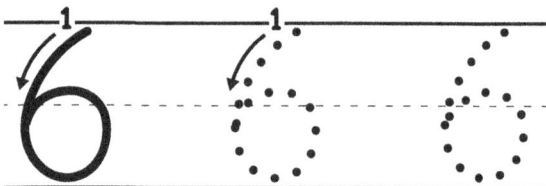

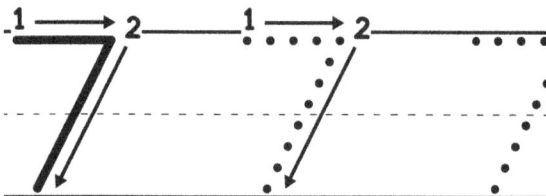

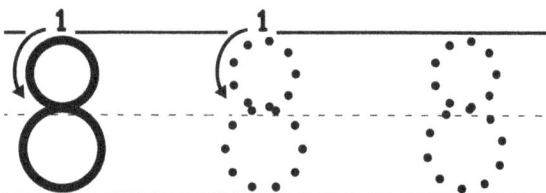

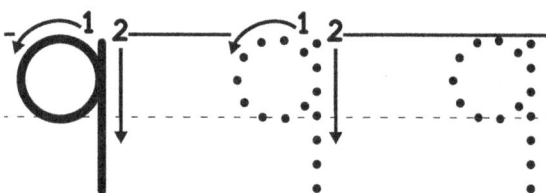

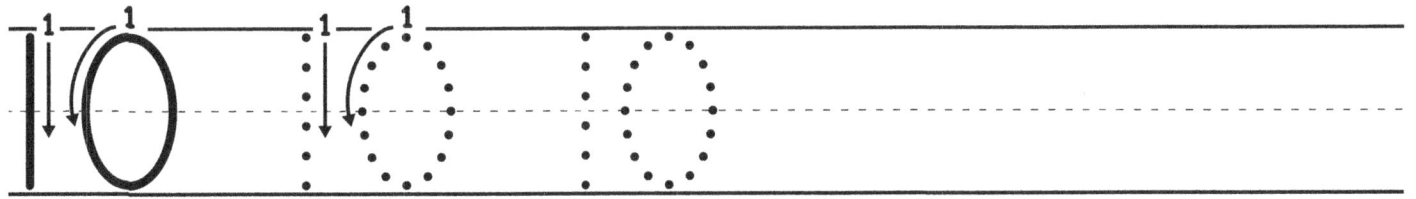

Fill out an ear tag for your cows! Don't forget to put your brand, initials, or a number in the middle.

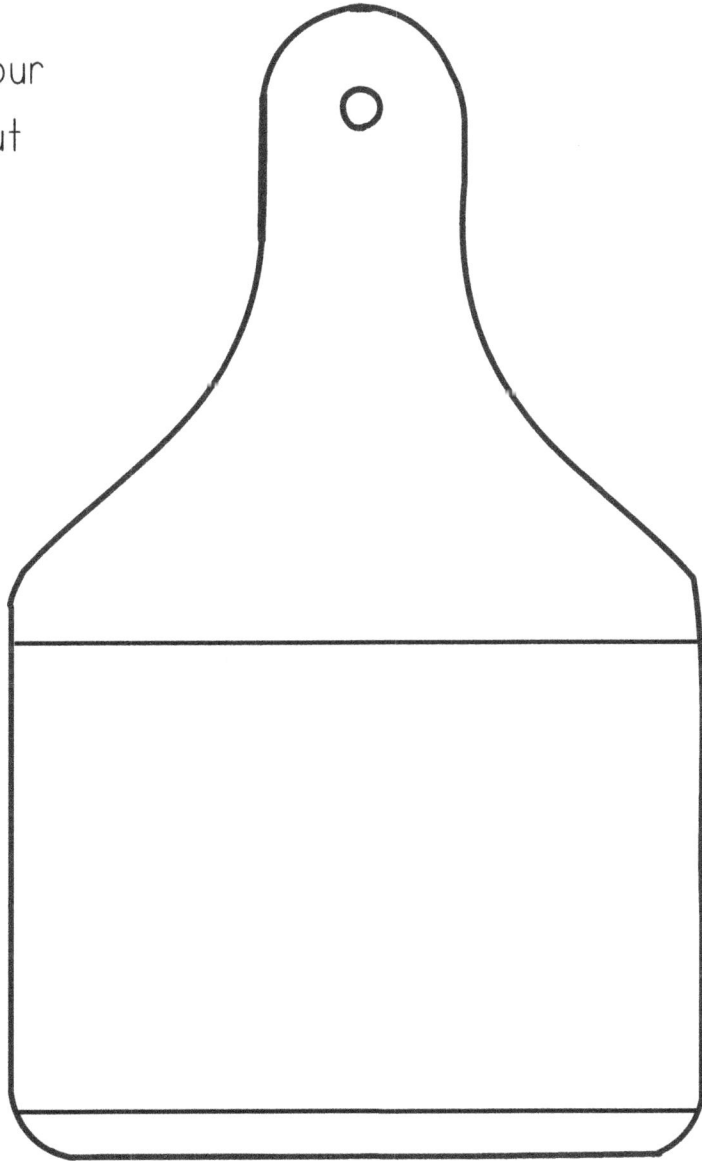

NAME:

Phone
Number:

REPLACEMENT HEIFERS

Rancher Danae decided to keep six of her heifer calves this year to replace older cows that can't have calves anymore. Can you number the ear tags for her? This year's replacement heifers will have pink tags.

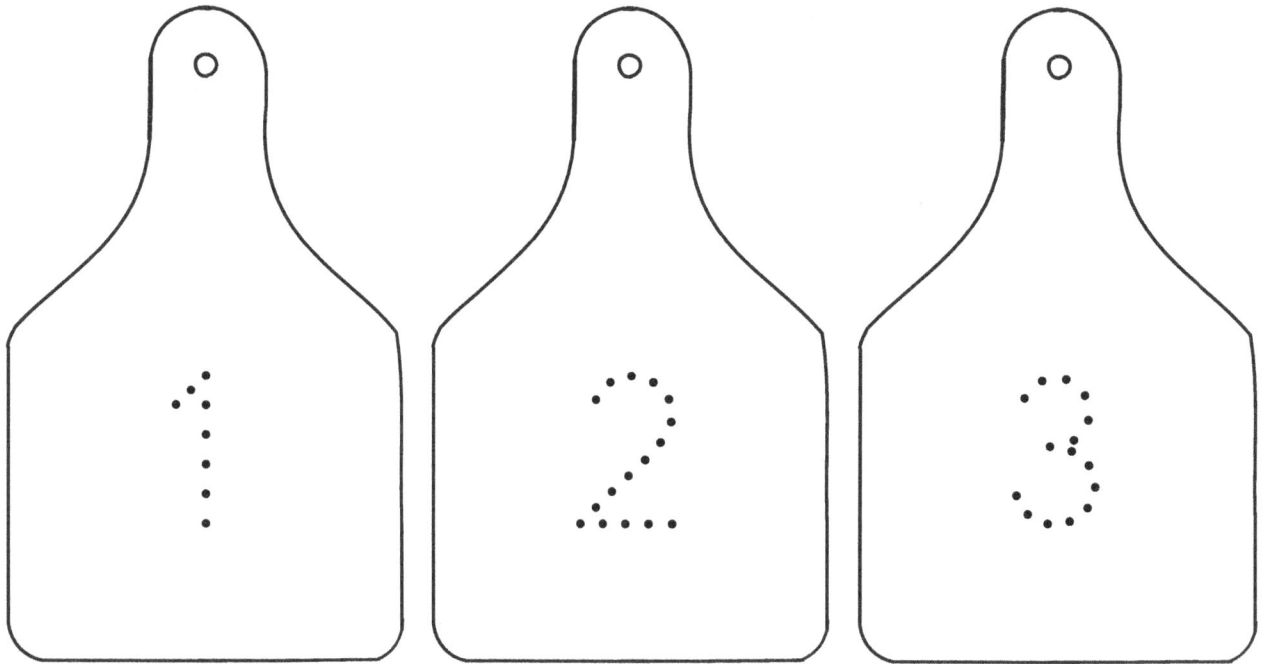

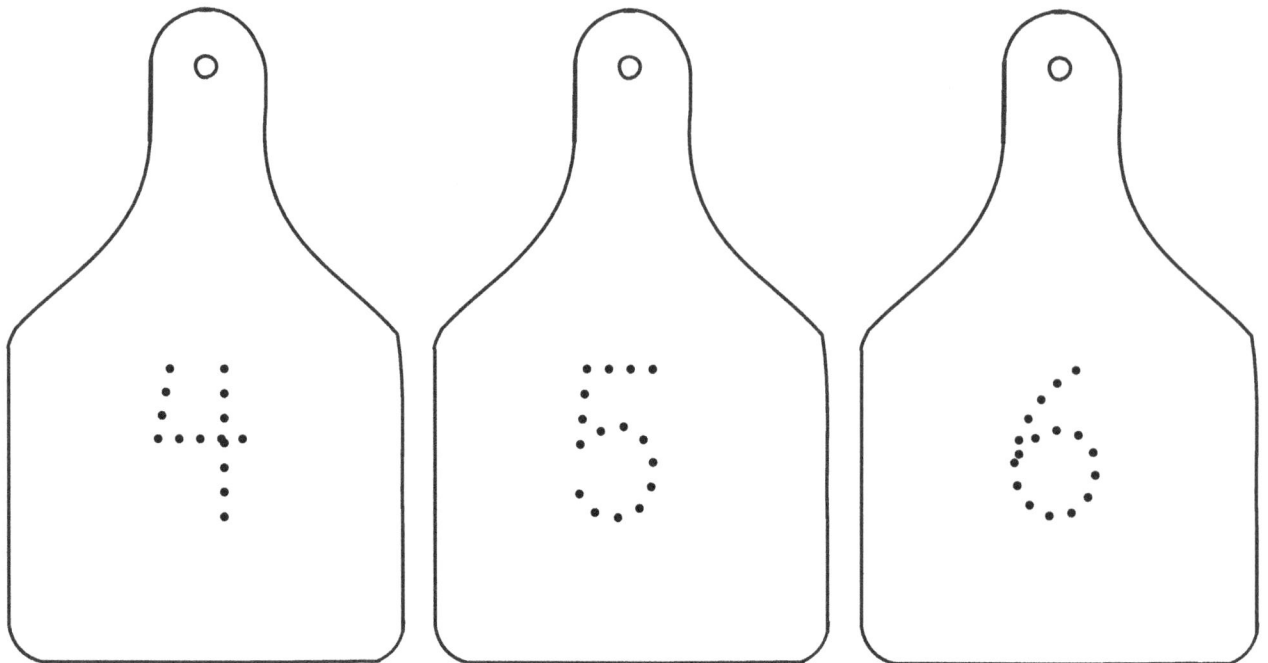

Color the Boot!

Robyn, the boot maker, needs help picking out the right colors of leather for the customer. Can you color the spaces to help her?

BLUE

BROWN

BLACK

GREEN

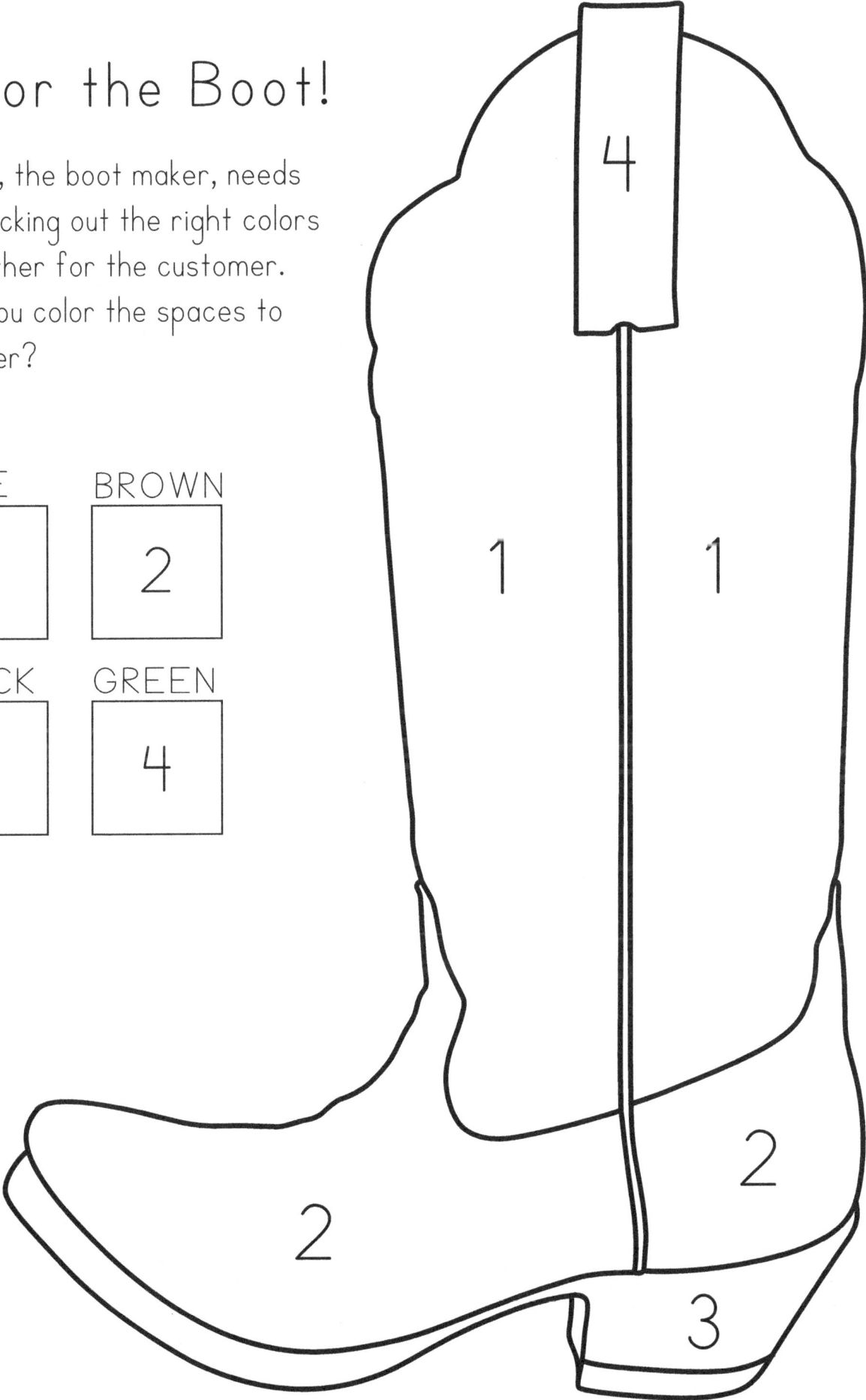

RAIN GAUGE RECORD

It rained last night! The cowboy needs to check his rain gauges to tell his boss how much rain they got during the storm.

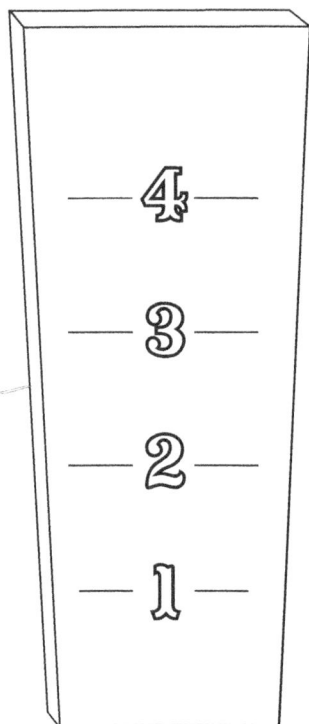

NORTH PASTURE GAUGE

It rained three inches in the north pasture. Can you fill in the rain gauge to show three inches of rain?

SOUTH PASTURE GAUGE

It rained one inch in the south pasture. Can you fill in the rain gauge to show one inch of rain?

There is a number at the beginning of each row. Can you color the same number of objects in the rows?

3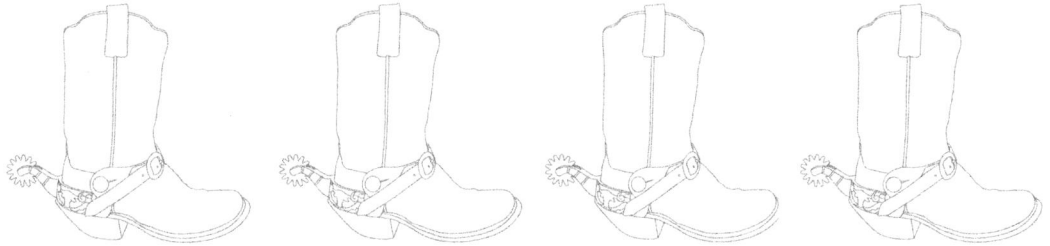

5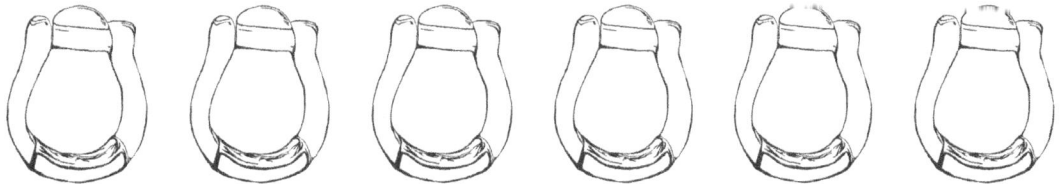

2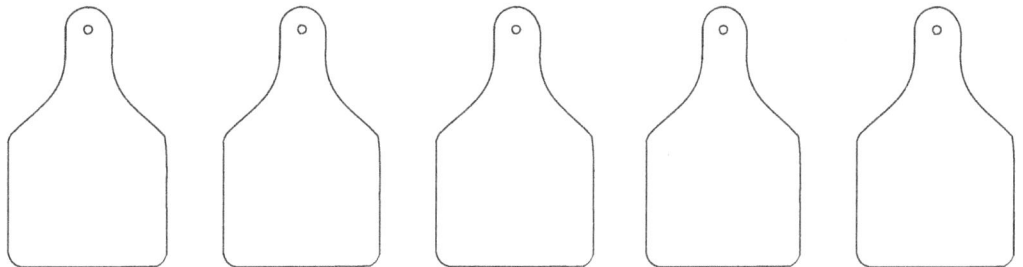

7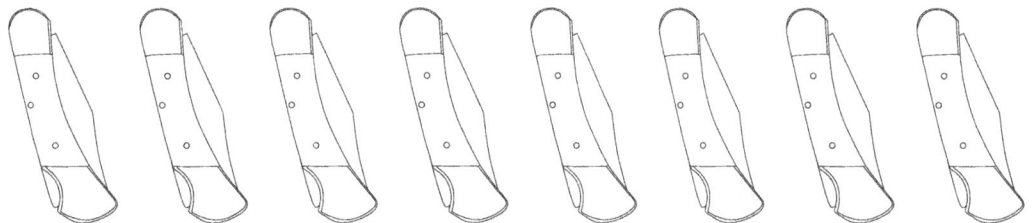

It's time to feed the horses! Apple, the horse, needs water. Can you color her circle water tank blue to fill it up?

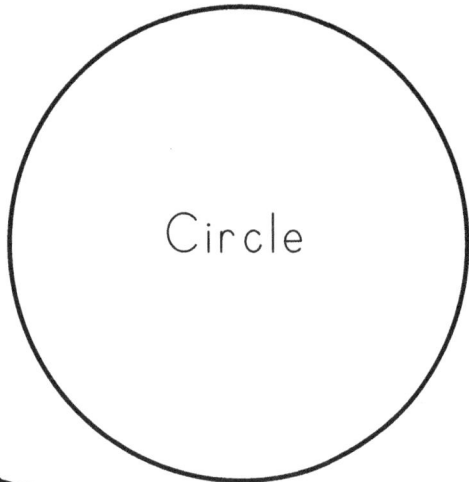

Circle

Now let's throw her some hay! Hay bales are rectangles. Can you color Apple's hay yellow?

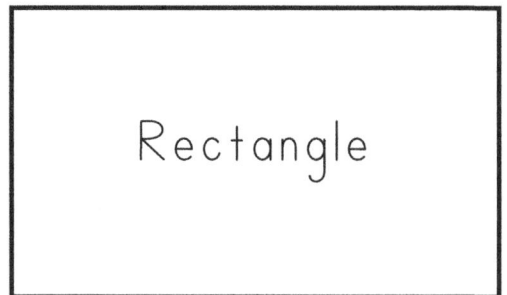

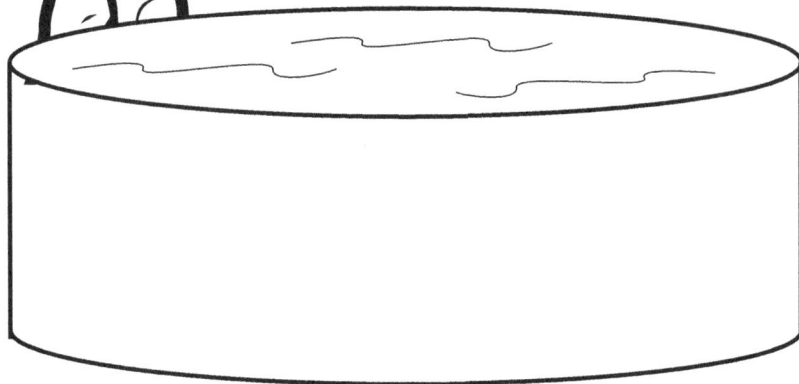

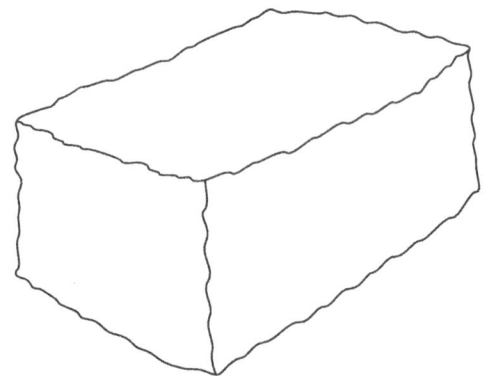

Rectangle

Great job feeding the horse!

Color the Knife!

Cowboy Coy got a new pocket knife today! He wants to color this knife to look just like his new one. Can you help him?

YELLOW

GRAY

RED

BLACK

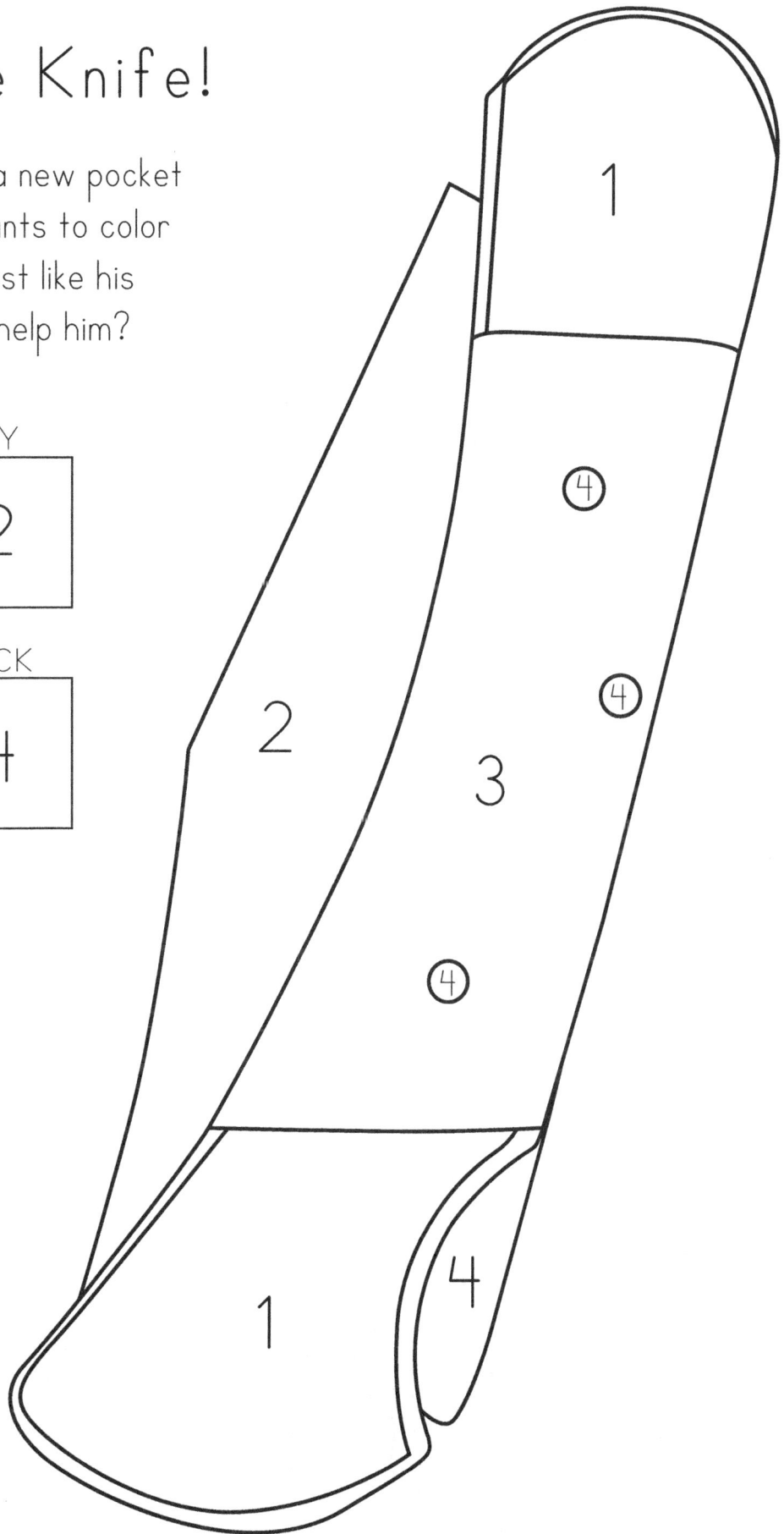

COUNT THE SALT BLOCKS

The cowboy needs help counting his blocks before he goes to feed. Can you trace the numbers and help him count the salt blocks?

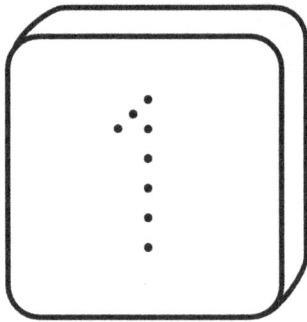

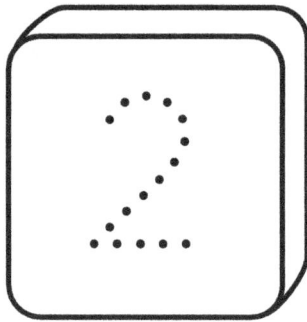

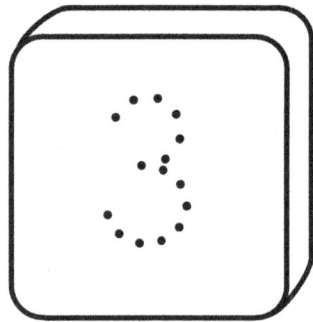

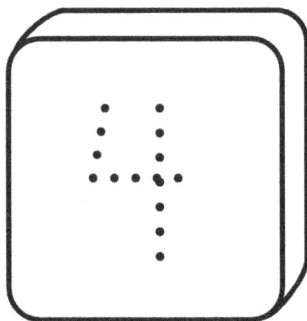

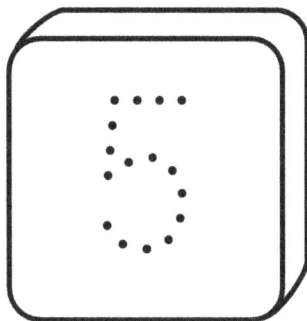

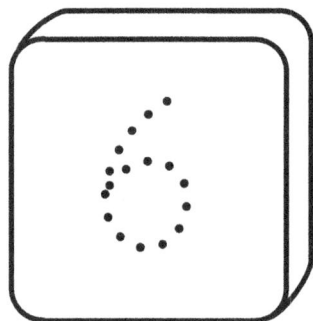

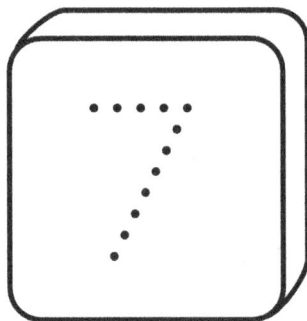

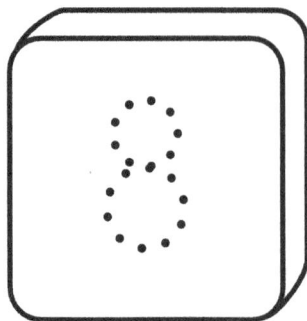

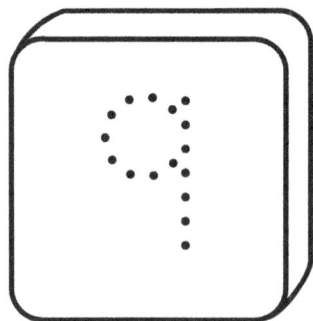

1 2 3

4 5 6

7 8 9

10

Can you help the leather maker finish the leggings by drawing a square pocket on one side and a rectangle pocket on the other side? Then you can color the leggings!

A bear got into the cowboy's teepee! His things are scattered everywhere. Can you help him color and count his belongings?

_____ Pants
_____ Ear Tags
_____ Cinches

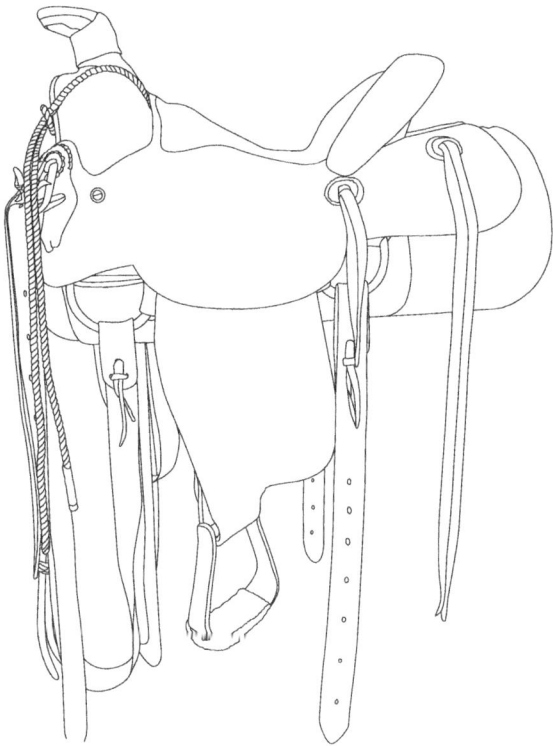

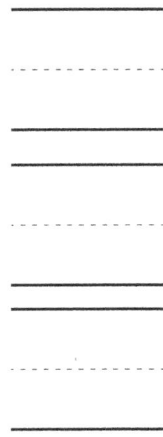

_____ Knives
_____ Boots
_____ Saddle

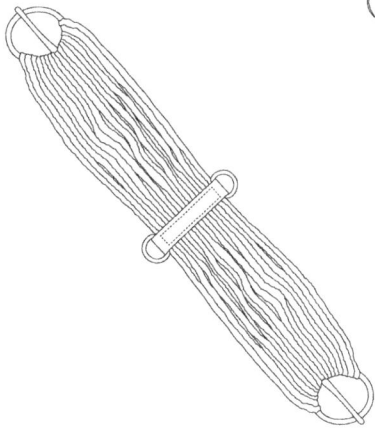

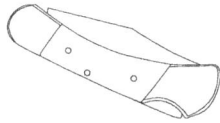

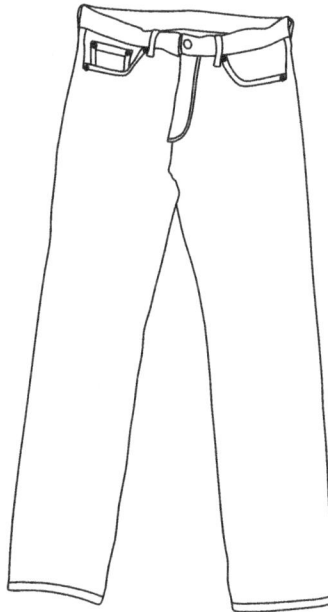

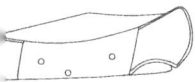

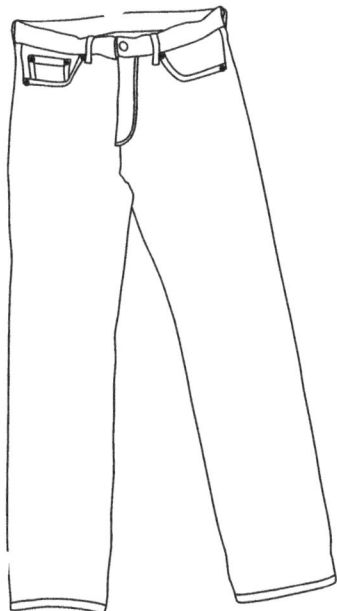

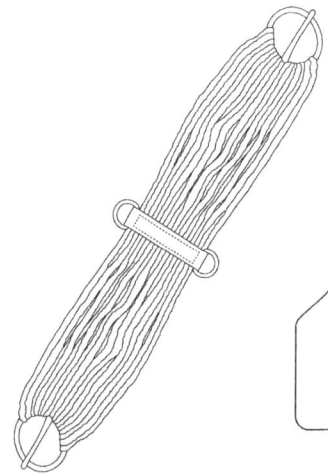

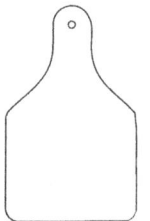

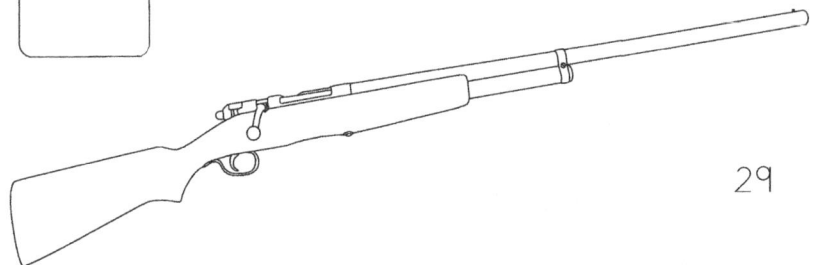

29

DOCTORING YEARLINGS

Flint and Rem need to doctor four sick cattle. They need to take the correct number of blue ear tags with them. Can you number and color four of the tags for them?

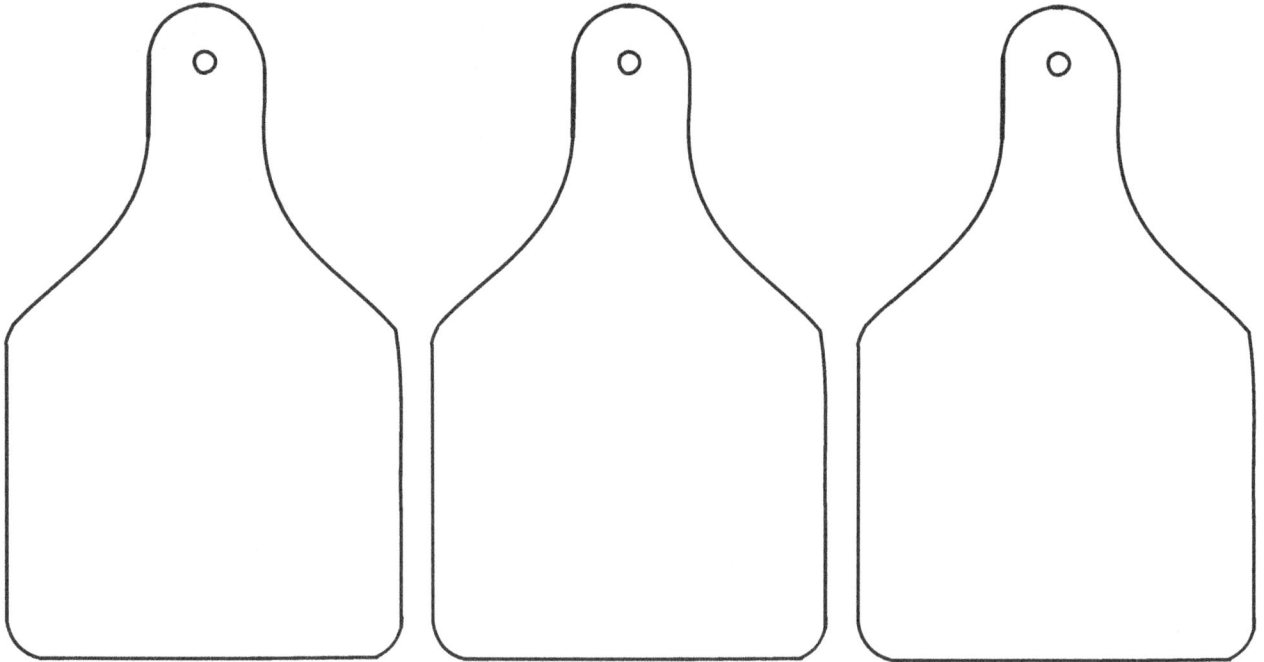

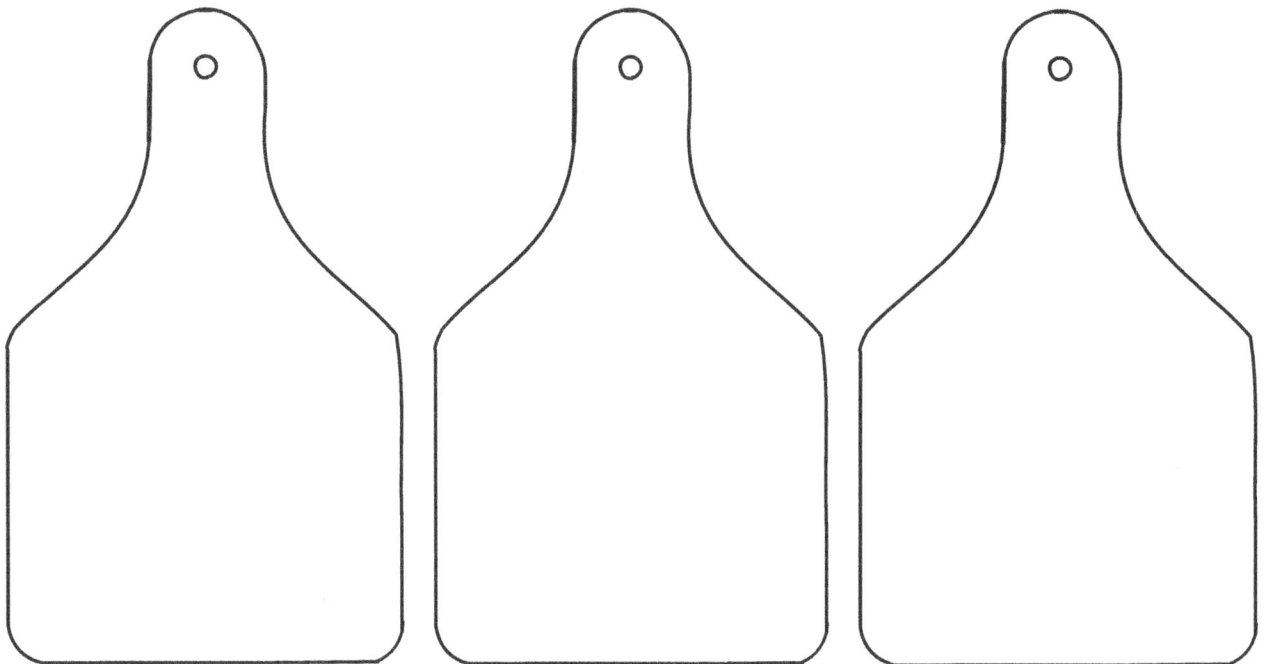

BRAND & COLOR THE COW

This cow needs to be branded. Can you brand her with your initials?

Great job! This is a Hereford cow. That means she's red. Two of her legs are red and two are white.

Dad is going to feed the cows! If he is going to feed one pasture this morning and one pasture after lunch, how many pastures will he feed in total?

_____   +   _____   =   _____

LOAD SALT AND MINERAL

The rancher needs to load four salt blocks and two mineral blocks on his pickup. Can you draw the correct number of squares on the back of the pickup and then color two of them brown to look like mineral blocks?

PAIR UP THE COWS

Can you help the cows find their calves?

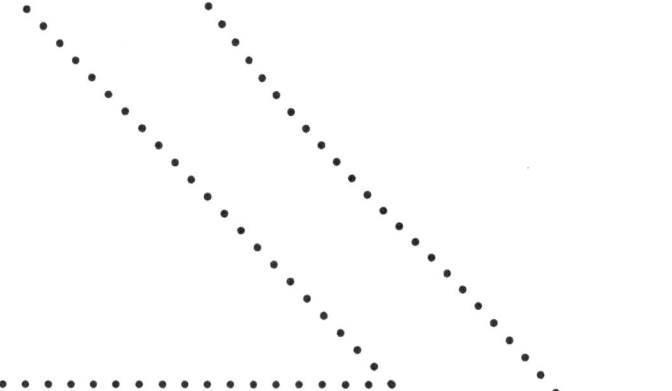

FILL THE GLASSES

Amy is cooking lunch for the cowboys and cowgirls who are branding calves today. She needs help filling the glasses with tea and lemonade.
Can you color the first glass brown for tea?
Then, color the second glass yellow for lemonade.

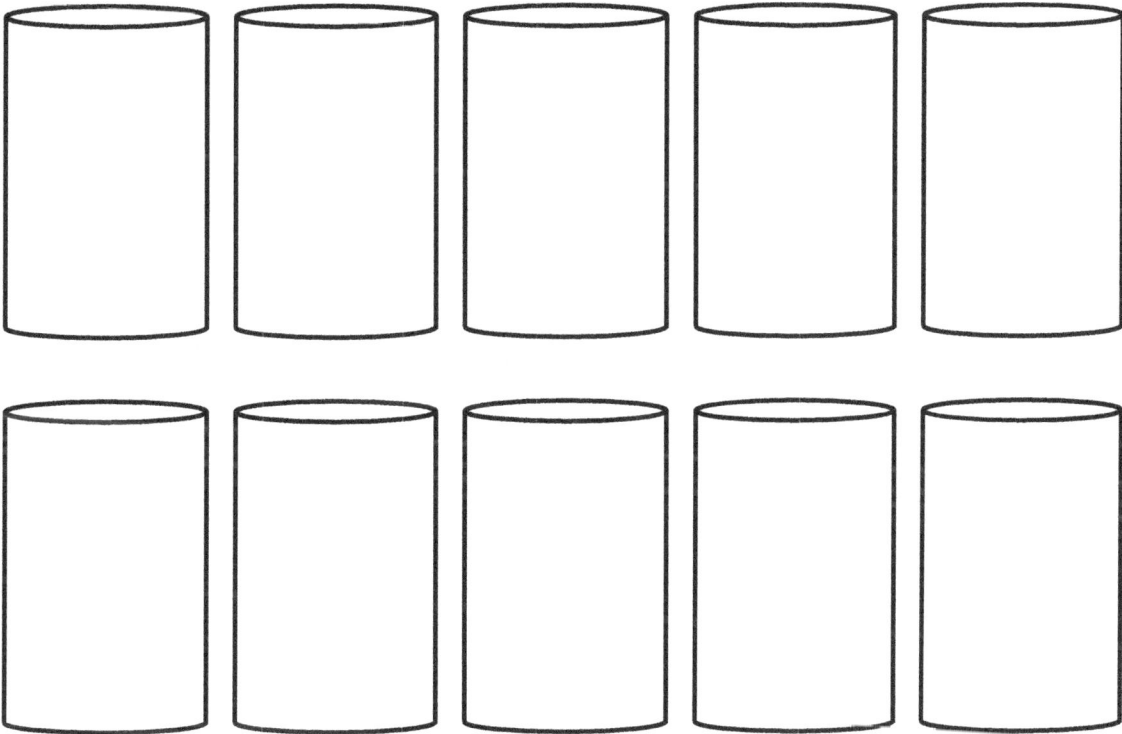

Finish coloring the glasses in the same pattern until they are all filled.

How many glasses
did you fill with tea?

_____

- - - - - - - - - - - -

_____

How many glasses did
you fill with lemonade?

_____

- - - - - - - - - - - -

_____

A remuda is a group of broke horses that cowboys use daily. Pa needs to rope four horses for his grandkids: Tealie, Reagan, Coy, and Abby. Can you help Pa rope four horses? Draw a line from Pa's hand to the four horses you think he should catch!

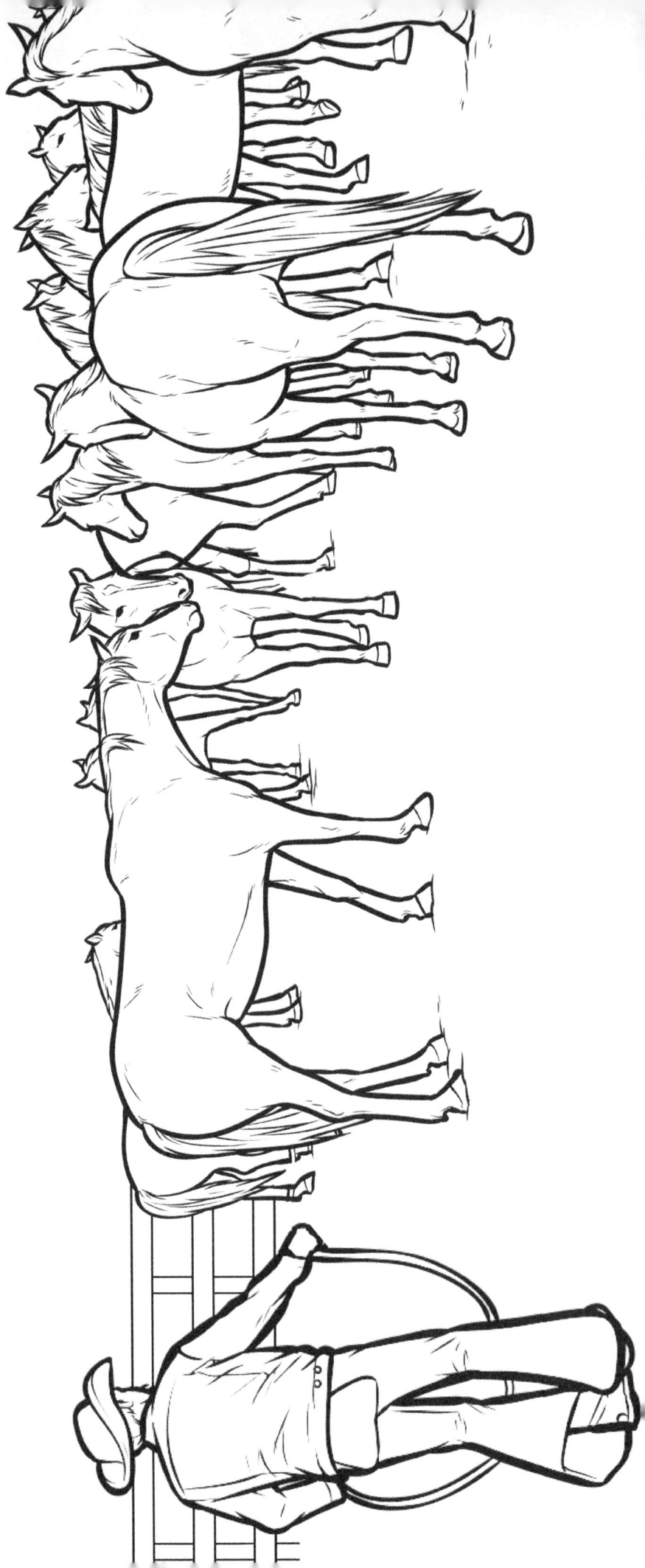

DAYS OF THE WEEK

There are seven days in a week. Can you fill in the missing numbers in the week?

Sunday	Monday	Tuesday	Wednesday	Thursday	Friday	Saturday
1		3		5		7

If the rancher feeds on Monday, Wednesday, and Friday, how many days per week does he feed?

7   3   4   2

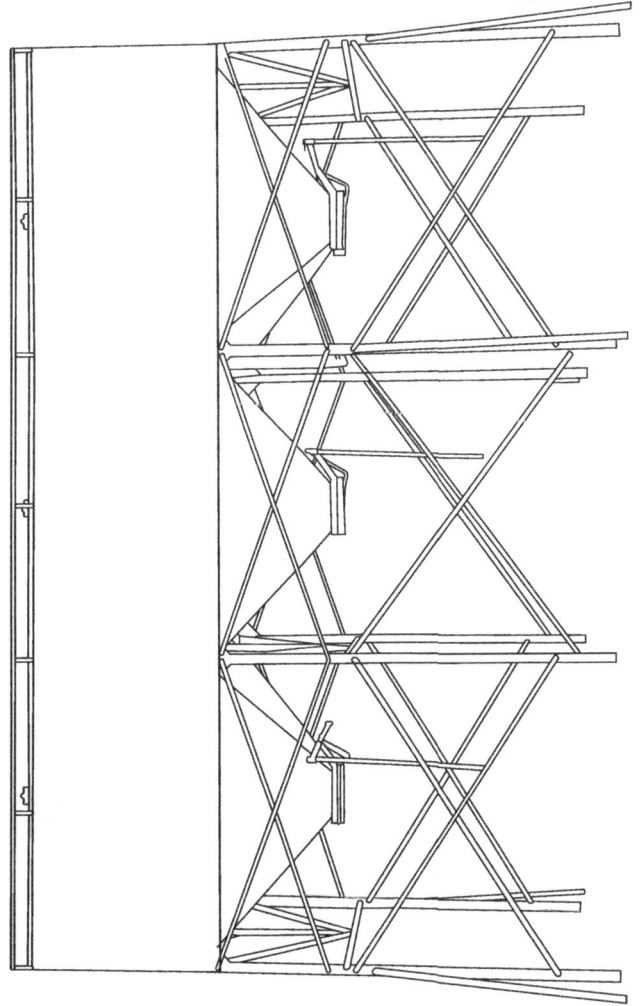

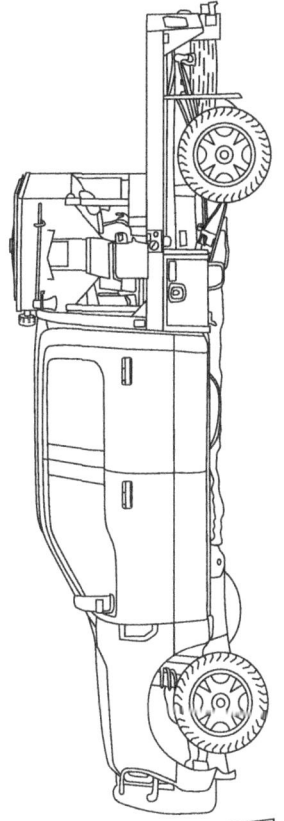

37

FIND THE COWS!

Cowboy Brady needs help finding a missing pair. Can you trace the lines to help him?

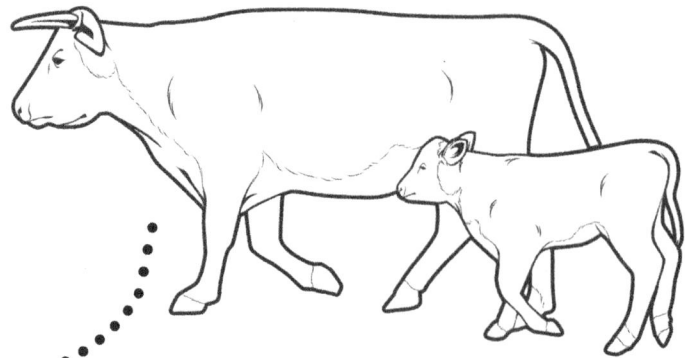

COUNT AND COLOR!

How many TIRES
can you count? ____

How many CACTI
can you count? ____

How many COWS
can you count? ____

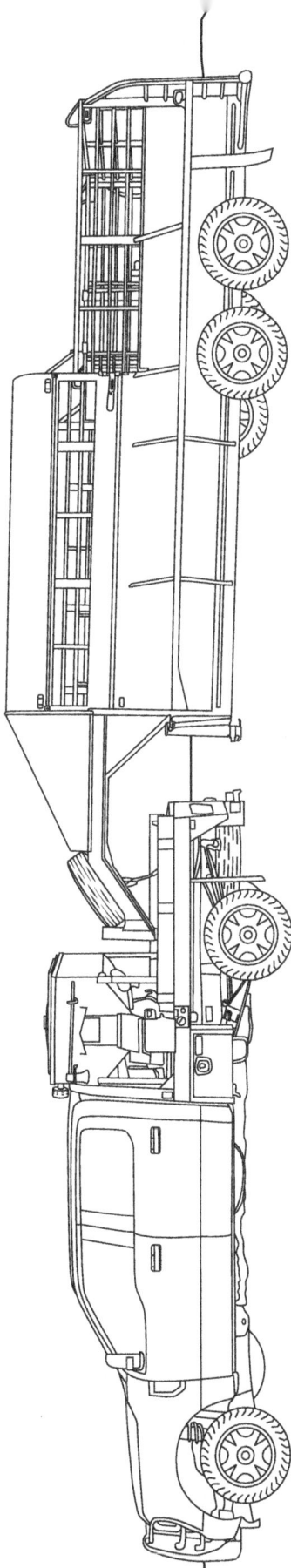
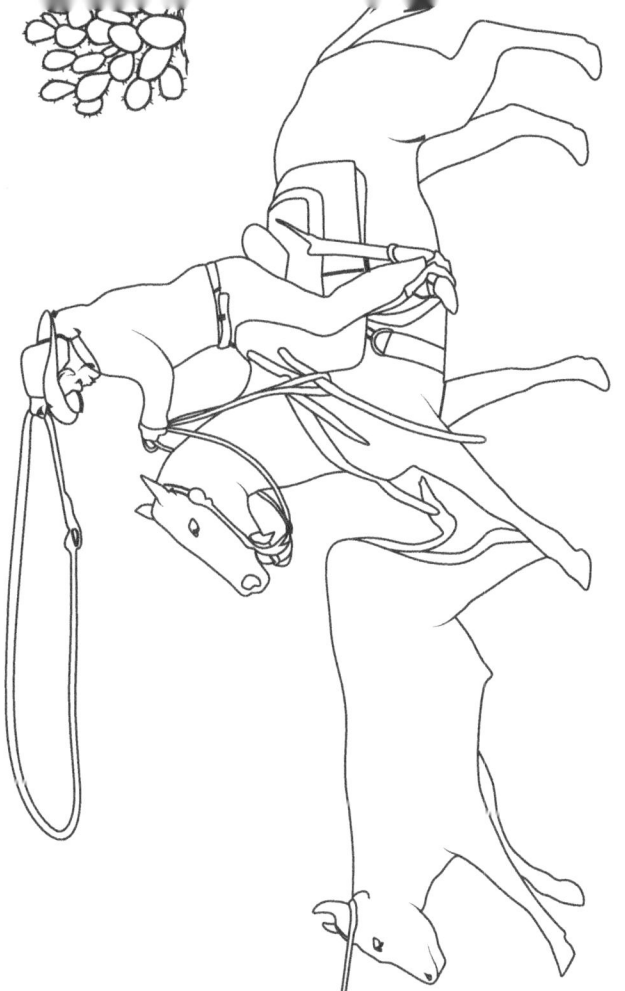

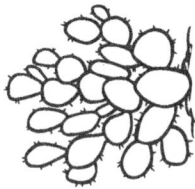

39

COUNT THE COWS

The rancher needs to count the cows as he feeds. Can you help him by drawing a line from the cows in the order they should go in?

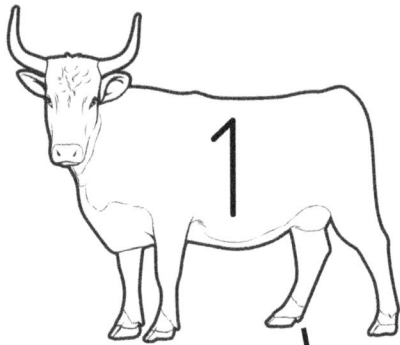

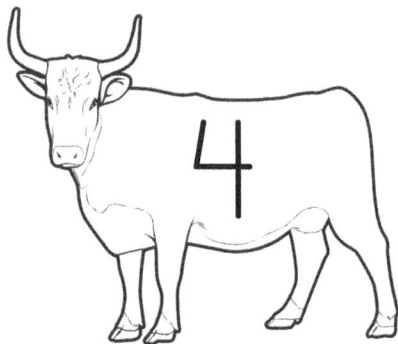

Count the horns while you color!   _____

How many horns did you count?   _____

After a long day of chasing critters Penny, the dog, is ready for a nap. Can you draw a dog bed in the shape of a rectangle so she can lie down?

Can you color the object that is different in each row?

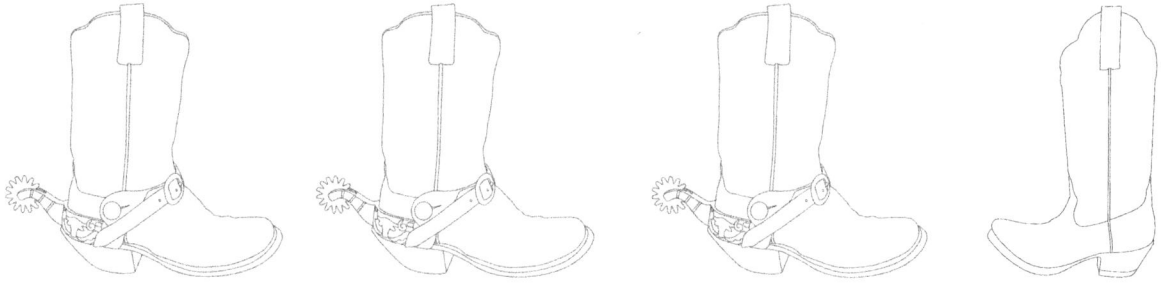

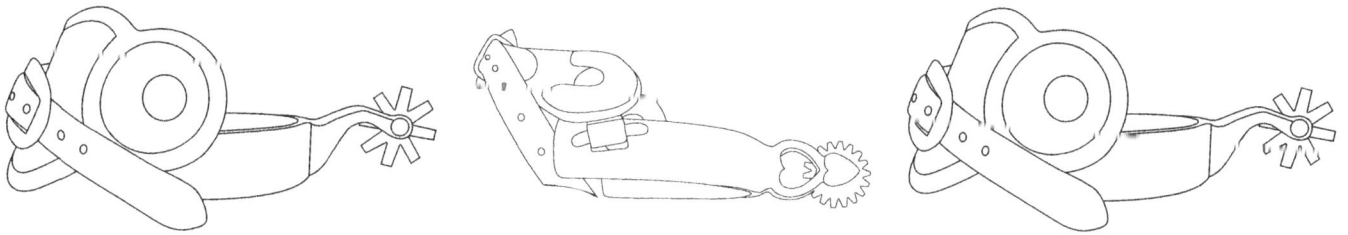

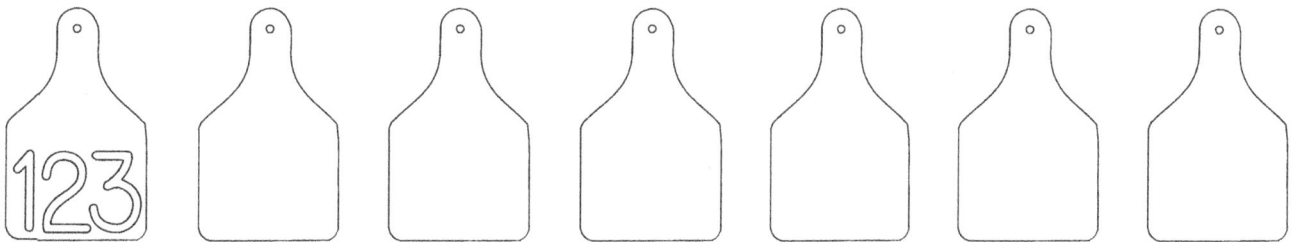

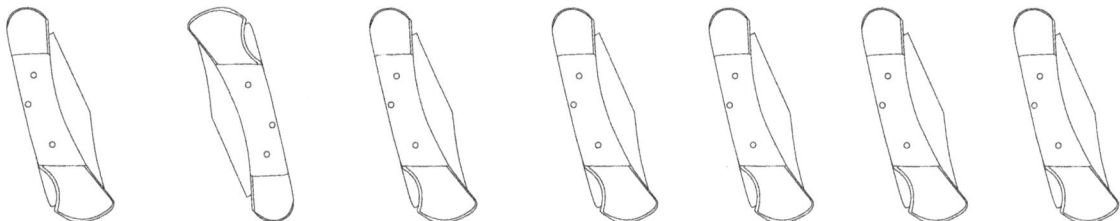

ROUND BALES

The rancher got in a load of hay bales. The ends of round bales are circles. Can you count the hay bales by practicing writing your numbers?

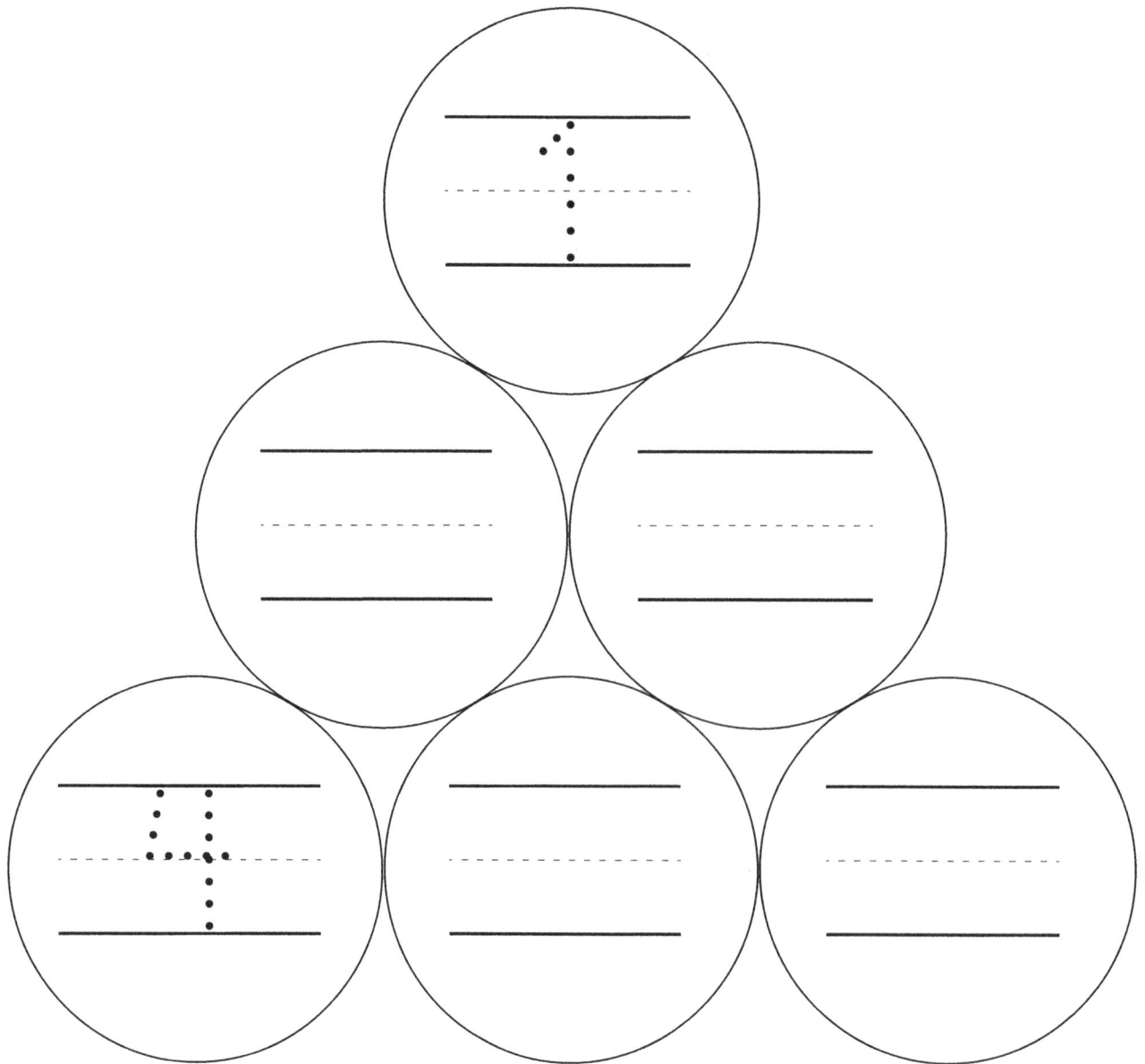

Great Job! Now color the hay bales yellow and green.

Connect the dots to finish the picture of the dog!

LUCY

1
2
3
4
5
6
7
8
9
10

What is the dog's name?

ROPE THE DIFFERENT ANIMAL

Oh, no! Someone left the gates open, and the livestock got mixed up. Can you rope or circle the animal that doesn't belong in each pen?

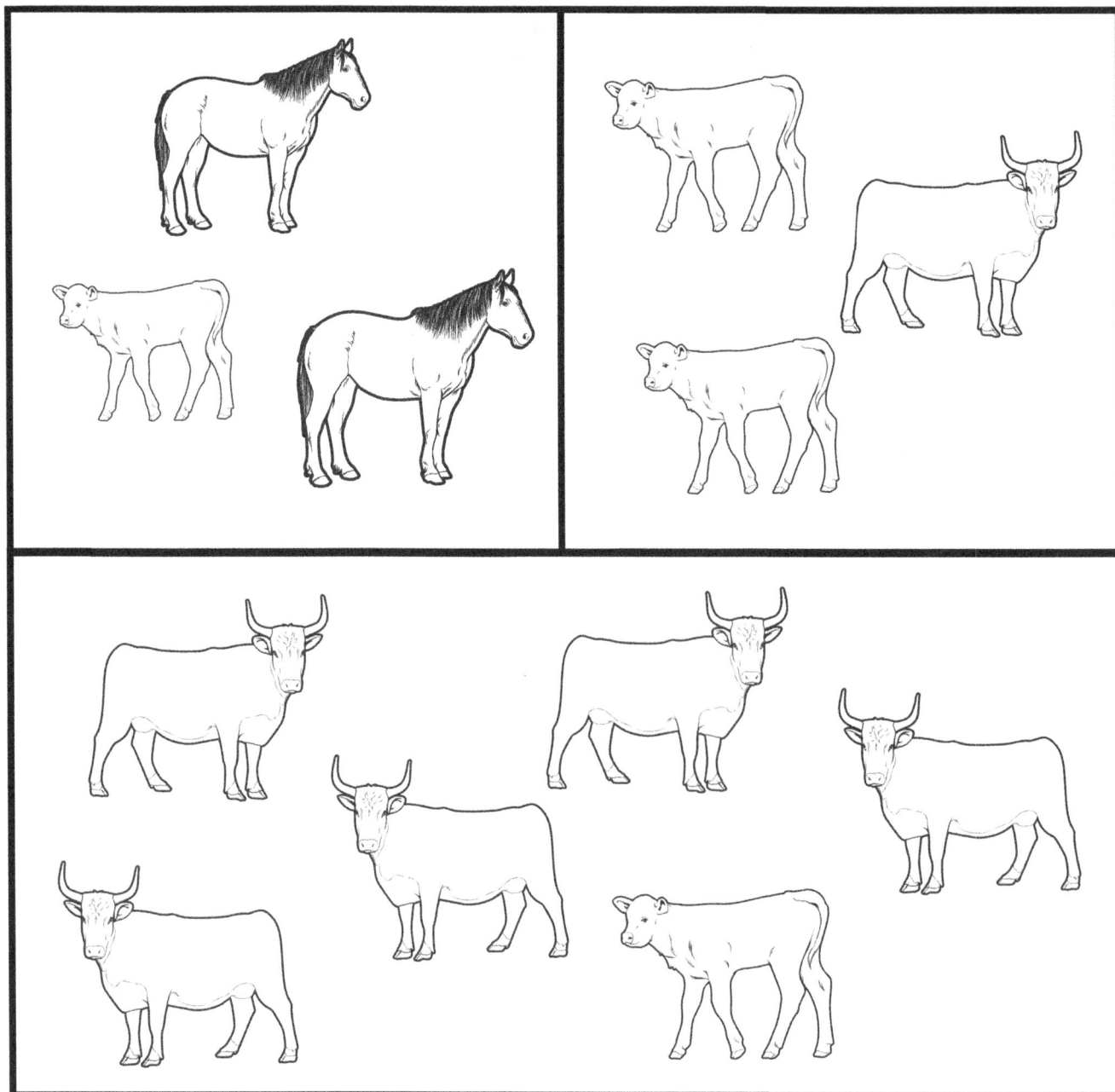

_____ How many CALVES can you count?

_____ How many HORSES can you count?

46

MISSING HORSE MYSTERY!

There should be three horses in the pens, but there are only two! Apple and Rooster are in the pens, but Hilldo is missing. Solve the puzzles to help the cowboy find his missing horse. Can you trace the horse tracks from the gate to see which direction Hilldo went?

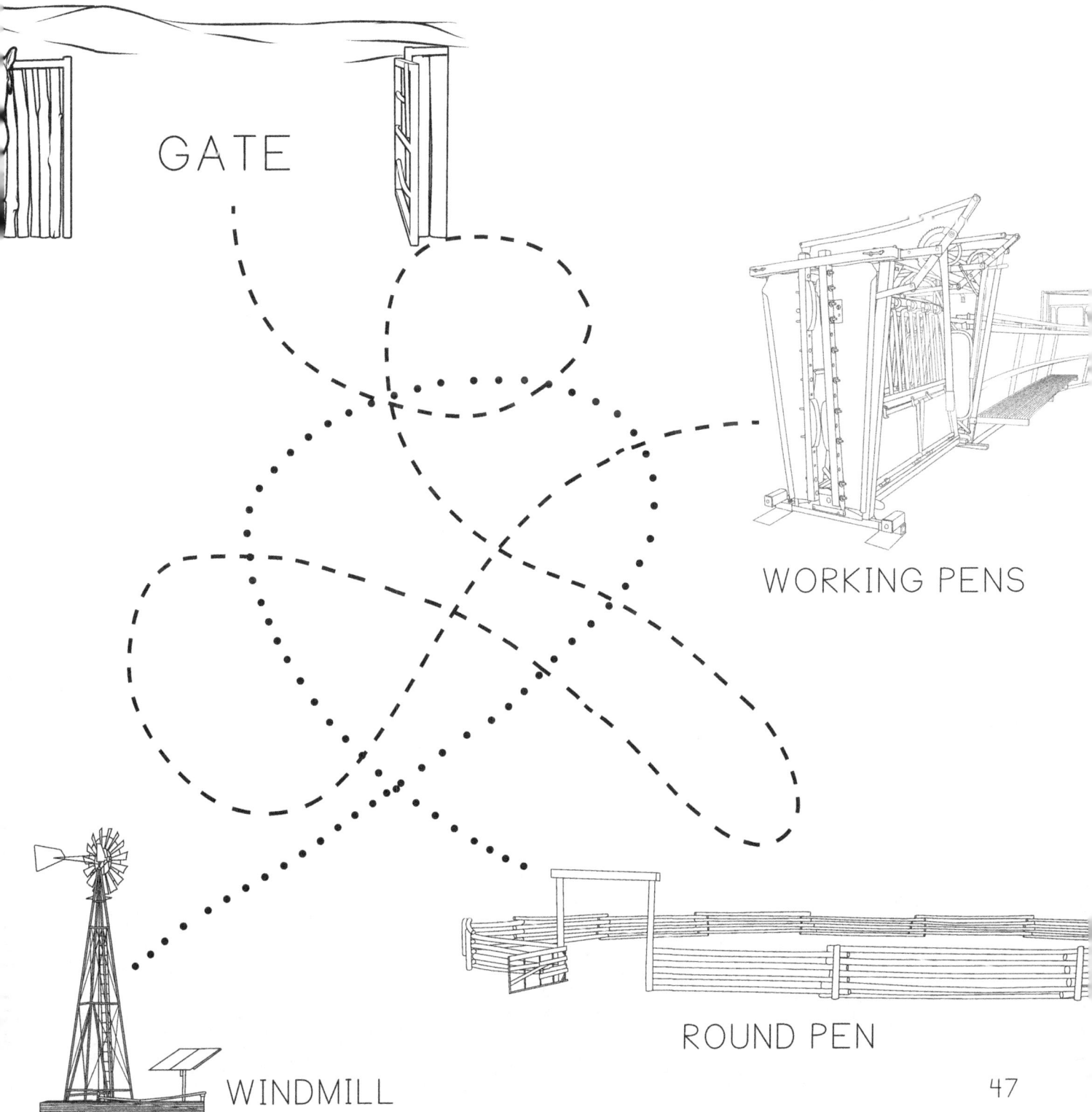

GATE

WORKING PENS

ROUND PEN

WINDMILL

HE WENT TO THE PENS!

The cowboy crew is helping the rancher wean his calves in the pens at headquarters today. They use the processing chute to work the cows during weaning. Hilldo must have heard the cows mooing at the pens and followed the noise.

There are the cowboys in the pens, but where is Hilldo? Something looks out of place and doesn't belong in the middle of all the livestock. Can you circle the thing that doesn't belong?

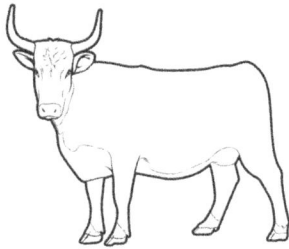

THE FEED PICKUP!

Silly Hilldo. The rancher set extra stirrups on the back of his pickup this morning. Hilldo must have wanted a snack from the feed pickup and knocked a stirrup off while he was eating. Stirrups come as sets or pairs. That means there are two of them. Can you circle all of the sets or pairs of stirrups?

How many pairs
of stirrups did
you circle?

FIVE PAIRS OF STIRRUPS!

Can you count the prints in each row below to see which path from the feed pickup has only five horseshoes? Write the number of horseshoes in each line on the back of the pickup. The path with five horseshoe prints is where Hilldo went next!

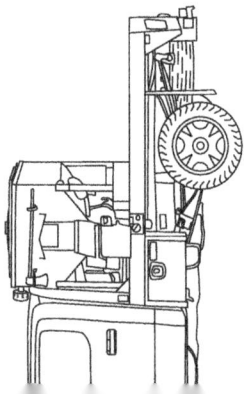

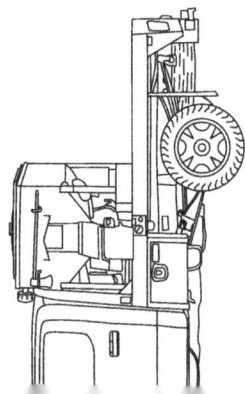

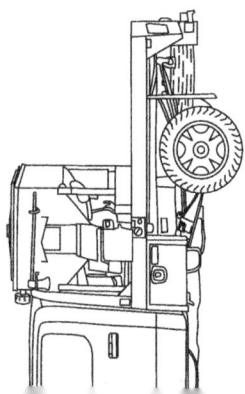

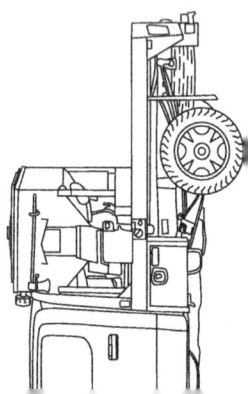

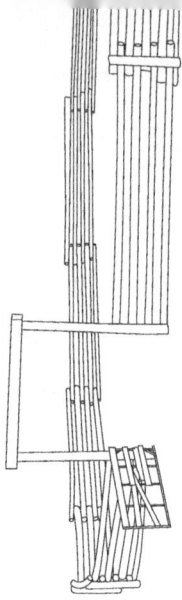

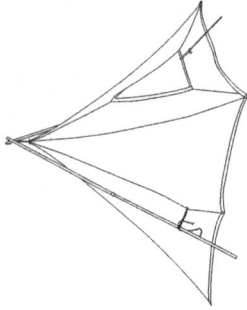

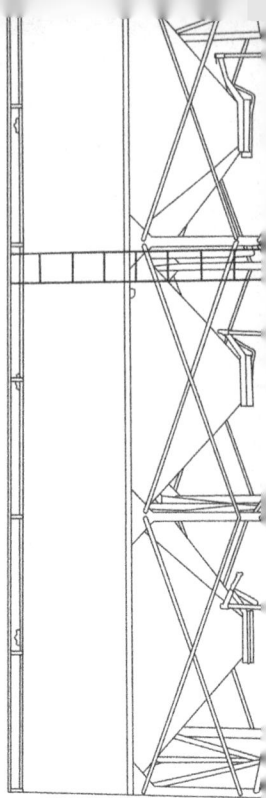

YOU FOUND HILLDO!

He was next to the overhead cake bins. He must have eaten all of the feed on the back of the pickup and then gone to look for more. He knew there would be some split on the ground near the cake bins.

For fun, let's count the pieces of feed on the ground in front of Hilldo!

_____ _ _ _ _ _ _____

1 2 3 4 5 6 7 8 9 10

Great job! You can count to ten!

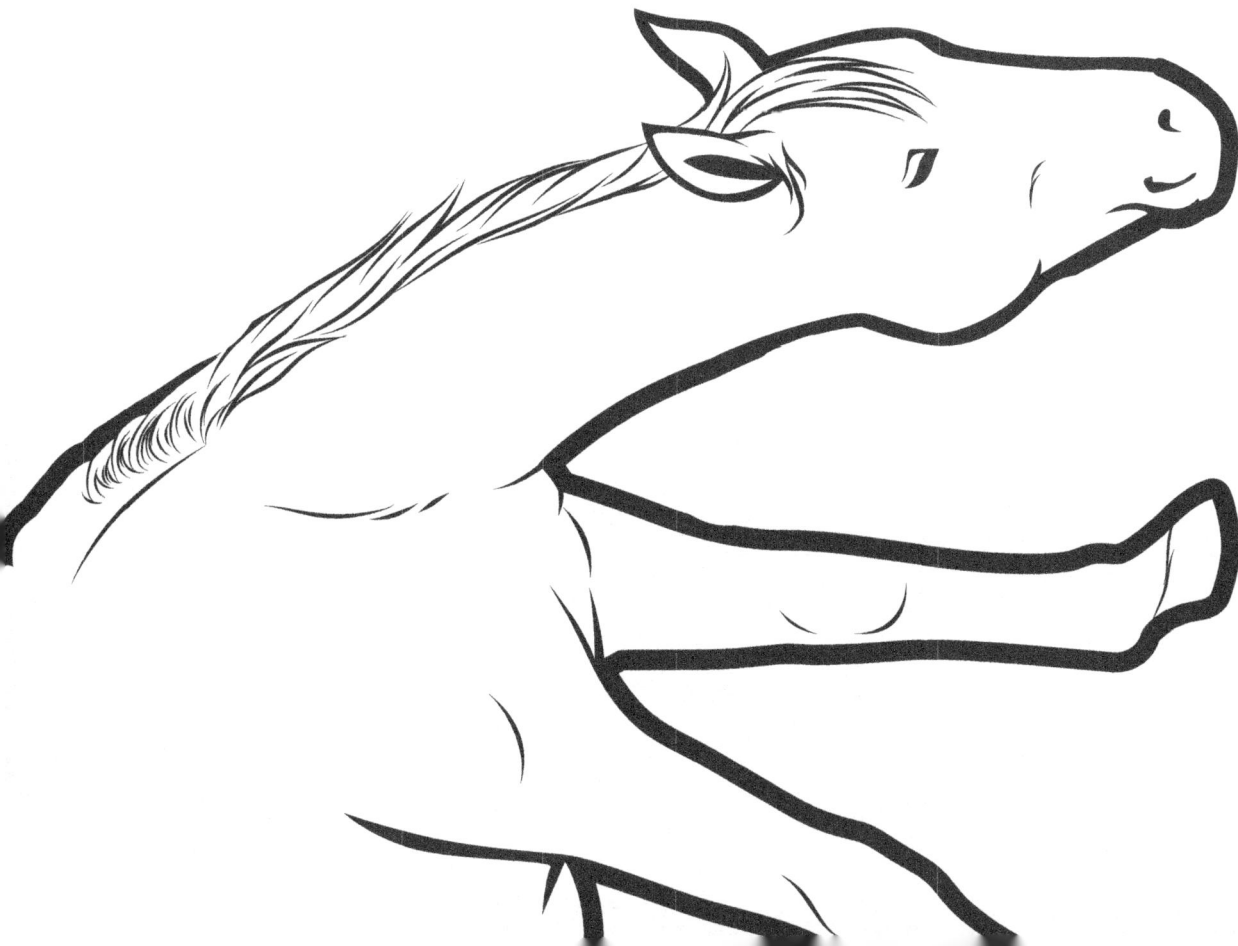

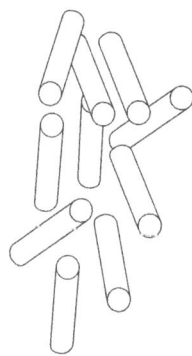

Made in the USA
Las Vegas, NV
16 March 2025